Tikitikiatongi Point
Spray Rock
Matapihi Point
Te Whau Point
Korakorahi Point
Woodlands Bay
Waiheke Point
Motuputuputu Point
Garden Cove
Cactus Bay
Stony Batter
215
Hooks Bay
Anita Bay
Kauri Point
Waiheke
Huse Bay
Garden Bay
Waiti Bay
Man o' War Bay
Rangitawhiri Point
Wairere Bay
Karipaka Bay
Waikopou Bay
WHITE COTTAGES
PUKE
RANGE
ISLAND
Awakiriapa Bay
Taniwhanui Point
Pakihi Point
Days Bay
Cowes Bay
Eel Rock
Arran Bay
Silver Bay
Pasadena Bay
Connells Bay
Finger Point
Waikorariki Bay
WAIHEKE CHANNEL
PAKATOA ISLAND
Frenchmans Cap (16)
ROTOROA ISLAND
Ruthe Passage
Maunganui
230
Awaawaroa Bay
Te Matuku Bay
Omiha
Sunday Rock
Passage Rock (12)
Kauri Point
Ponui
Ponui Head
WIND GENERATORS
Te Kawau Bay
PONUI ISLAND
(Chamberlins Island)

FOR MARTI AND GERRARD; MY MOTHER MARGARET AND MY FATHER TERRY; FOR DENIS AND MY FAMILY – THANK YOU.

First published in 2017 by Beatnik Publishing

Editor: Janet McAllister

Art Direction & Design: Sally Greer

Cover Painting (detail): *A Map of Church Bay and Terroir*, Denis O'Connor, 2014, Private Collection, slate and acrylic
Inside Cover Map: *Tamaki Strait and Approaches including Waiheke Island*, 2012, Land Information New Zealand
Painting on page 198 *Mother Nature*, 1963, Joseph Alach

Printed and bound in China.

ISBN 978-0-9941383-3-0

Beatnik Publishing

PO Box 8276
Auckland 1150
New Zealand

www.beatnikpublishing.com

WAIHEKE ISLAND

—

A WORLD OF WINE
THE PEOPLE BEHIND THE LABELS

CLARE DUNLEAVY

PORTRAITS BY
MARTI FREIDLANDER

CONTENTS:

FOREWORD: **BOB CAMPBELL**:7

INTRODUCTION: **CLARE DUNLEAVY**:8

DEDICATION: **MARTI FRIEDLANDER**:10

GOLDWATER:12
KIM & JEANETTE GOLDWATER
WINE AND GENEROSITY

STONYRIDGE:18
STEPHEN WHITE
WINE AND FORESIGHT

TE MOTU:24
PAUL DUNLEAVY
WINE AND INHERITANCE

JOHN DUNLEAVY
WINE AND INVENTION

FENTON:34
BARRY & MEG FENTON
DESTINATION WINE

MUDBRICK:40
ROBYN JONES
WINE AND ROMANCE

NICK JONES
WINE EQUATIONS

MAN O' WAR:52
MATT ALLAN
WINE FOR LIFE

HERB FRIEDLI:58
THE BAREFOOT CONNOISSEUR

MIRO:64
CAT VOSPER & BARNETT BOND
WINE AND THEATRE

OBSIDIAN:70
MIKE WOOD
WINE AND MIGRATION

PASSAGE ROCK:76
DAVID & VERONIKA
EVANS-GANDER
WINE PIONEERS

TE WHAU:82
TONY FORSYTH
1 + 1 = WINE

KENNEDY POINT:88
NEAL KUNIMURA,
SUSAN MCCARTHY &
ELIZABETH ZHONG
A CULTURAL FERMENT
—

AWAROA:94
STEVE POLETTI
WINE AND COSMOLOGY
—

CABLE BAY:100
LOUKAS & CAROLINE PETROU
A WINE ODYSSEY
—

DESTINY BAY:106
MIKE, ANNE & SEAN SPRATT
WINE AND OPPORTUNITY
—

MICHAEL'S:112
LOUISE THOMAS &
MICHAEL HOLDEN
WINE FLIGHTS
—

JURASSIC RIDGE:118
LANCE BLUMHARDT
WINE AND GEOLOGY
—

HAY PADDOCK:124
CHRIS CANNING
WINE AND THE CHAMELEON
—

WOODSIDE HILL:130
DAVID & HELÉN GUNSON
WINE: AN OUTDOOR PURSUIT
—

FRENCHMANS HILL:136
LUC DESBONNETS
LA VIE ET LE VIN
—

PEACOCK SKY:142
CONNIE FESTA &
ROB MEREDITH
WINE AND PATIENCE
—

PODERI CRISCI:148
ANTONIO CRISCI
WINE AND LA DOLCE VITA
—

SOHO:154
RACHAEL CARTER
WINE AND CLOSURE
—

BATCH:160
CHRISTINA & CRAIG THOMAS
COOL CLIMATE WINE
—

GOLDIE: 166
NEILL CULLEY
WINE AND INDUSTRY

HEINRICH STORM
WINE AND PARTNERSHIP
—

ARAE:178
CHRIS & MIKE POLAND
FROM OBSERVERS
TO PARTICIPANTS
—

WILD ESTATE:184
KERYN & SIMON MATTHEWS
A TALE OF TWO ISLANDS
—

TANTALUS:190
CARRIE MENDELL &
CAMPBELL AITKEN
UNLOCKING THE SECRETS
OF WINE
—

CHRONOLOGY:197

FOREWORD
BOB CAMPBELL

—

The quality and character of wine is shaped by many things. Soil and climate get top billing as does the chosen grape variety or varieties. Trellising, pruning, leaf-plucking, fruit-thinning and a whole host of viticultural and winemaking techniques also make a contribution.

To truly appreciate the character of a wine it helps to understand the character of the person who chose to buy the land, select the grape varieties and trellising methods, decided when to begin harvesting, and was the ultimate decision maker from fermentation temperature to label design. Are they a risk-taker? Do they embrace tradition? What motivated them to become a wine producer in the first place?

The personalities behind the labels are as fascinating to me as the wines they produce. Clare Dunleavy clearly feels the same. For example, when I sip a glass of Te Whau "The Point," I think of its enthusiastic and deeply intelligent owner Tony Forsyth tackling a vineyard site so steep that many believed it could not be done. His wine tastes edgier and more profound for having met him. Clare helped me see Tony in a new light. My next glass of Te Whau will have even more depth as a result.

When I received a first draft of *Waiheke Island – A World of Wine, the people behind the labels*, I eagerly read the chapters featuring people I know well before turning to those about makers I know less well or not at all. Their stories add personality to a wine region that has an abundance of character and characters.

Waiheke wine will never taste the same again.

—

Bob Campbell, MW

INTRODUCTION

—

Waiheke Island lies in the Hauraki Gulf, just a short boat ride away from downtown Auckland. It's a small island with a large reputation for quality wine production. Our population hovers around 8000, and many of us lucky enough to call Waiheke home work in the wine and associated tourism industries.

I moved to Waiheke in 1993 after many years spent in Europe. While I was away, my family purchased land and planted grapes on what would become Te Motu Vineyard. There were just a handful of vineyards back then. As you travel through this book, you will be introduced to the bravely determined characters whose vision and goals have flourished and now flow into every glass of Waiheke wine poured here and around the world. They are our vineyard owners - the foundation upon which our local wine industry has been built. Some of them you will know or have the chance to meet by visiting our Cellar Doors or Restaurants. Many vineyard owners work beyond the public gaze. You will also meet a few winemakers who have an enduring commitment to our Waiheke wine industry. I have gathered them all together in this book to tell their stories. They represent many diverse cultures and professions and are a microcosm of the wider world we inhabit.

Our first vineyard was planted by the Gradiska family in Ostend, in the 1940's. They had a market garden selling fruit and vegetables and made their own wine. The painting on the page 198 is a detail of their vineyard by Joseph Alach after visiting his friend's home. This painting hangs above the fireplace in our home, a constant echo of the past and of the impressive progress of our wine industry.

Many voices are absent from this book – former vineyards such as Peninsula and View East, whose original owners are no longer producing wine. John Hawkesby's vineyard's tiny production is not for sale, but has often been gifted to our community to raise money and awareness for particular projects. Manuka Valley is another tiny family owned vineyard, producing Syrah. Glenora Estate's wines sell internationally. There have been many talented winemakers who have come and gone. And finally – there are hundreds of brilliant folk working on our vineyards and in our wineries, for whom I raise a glass in celebration of their skilled focus and contribution to the quality and reputation of our wines.

When I arrived on Waiheke, I worked on our vineyard: planting, field-grafting, pruning, harvesting, driving tractors, getting dirty, cold, burnt, wet, tired and curious. It was hard work that marked the beginning of my career in wine. I've worked in our Cellar Doors and Restaurants, studied wine, worked in the wine trade. I've become an educator, advocate and consumer of our wine. It has been my ambition for many years to introduce the people who searched for the right place in the world to plant their dream. They found Waiheke Island and created their future. I'm grateful to many people who have supported me in the writing of this book. The process has instilled in me a firm belief in this Maori proverb: *He aha te mea nui o te ao.* What is the most important thing in the world? *He tangata, he tangata, he tangata.* It is the people, it is the people, it is the people.

—
Clare Dunleavy

DEDICATION
MARTI FRIEDLANDER

—

I first met Marti Friedlander, one of New Zealand's most revered photographers, in the summer of 2002 while working as Sommelier at Te Whau restaurant and vineyard. Imagine how I felt that day when Marti told me that I had the most beautiful green eyes. "May I photograph your eyes?" she asked, in her unforgettable, richly textured voice.

That same year, my partner Denis O'Connor gifted me a new edition of Pioneers of New Zealand Wine, first published in 1964, written by Dick Scott, with photographs by Marti Friedlander. An idea began to form.

When I finally committed myself to this book, there was no question about the portrait photographer I dreamed of. It was Marti – and she said "yes". Into my book walked one of New Zealand's most adored and venerated portrait photographers.

Marti, her husband Gerrard and I spent two weeks visiting the people featured in this book. It was a privilege to witness Marti at work. She told me that "being a photographer is often very intrusive". I quietly observed her establishing trust and ease with her subjects so quickly, that in no time at all she had captured the moment. Gerrard is deeply passionate about wine and Marti was thrilled to share our project with him – "I loved his enjoyment of it – he really is in love with not only the drinking of it, but the making of it and he's extremely knowledgeable." "Gerrard and I enjoy wine as a sort of aperitif to life". They are both deeply curious about people, intellectually committed to culture and politics. In short – they are inspirational.

This book allowed me to step into the lives of Marti and Gerrard. Together with Denis and my family, they have been a constant source of generous and heartfelt support and encouragement. Marti and I emailed each other constantly and met for lunch often. Her loving encouragement inspired me to tackle this project; to overcome the enormity of undertaking something very real and lasting, to be responsible in my writing for each and every person I documented. To bear witness to the lives of our subjects, just as Marti spent her entire life doing so passionately, memorably, magnificently. We had finished editing my text and were moving into the design process when Marti died, aged 88, on the 14th November 2016, with her life-long love Gerrard by her side. I felt my heart break that day, for Gerrard and for all of us nestled within her warm embrace. She did read most of my stories, sending me wonderful notes in congratulations for what she felt sure would become a best-seller book. I wish she had been able to see our beautiful book as you see it. In a way, she is gazing out at us through the vision of her subjects, captured forever in a fleeting moment. Marti's big heart and wise words will live in me forever. I am deeply honoured to have Marti's portraits and her wise gaze enriching the pages of my book.

Marti told me "I love winemakers – they have a dream and they achieve it – you have to admire that sort of persistence and passion Clare – it requires such a lot of energy". Dearest Marti - I couldn't agree more.

KIM & JEANETTE GOLDWATER
GOLDWATER ESTATE

—

WINE AND GENEROSITY

Kim and Jeanette Goldwater are the godparents of wine on Waiheke Island. Throughout the year and a half of interviewing for this book, their names came up constantly in conversation. I began to feel that I was learning more about them from the wine community that has grown up around them, than from themselves, so fulsome is the island's respect and love for this eloquent and engaging couple. Still leading rich and full lives, they deserve all the admiration they attract.

Inspired by the wines they drank in Europe, in 1977 they bought a block of land in Ostend, and planted the first Vinifera grapes on the island. From this humble beginning, Goldwater Estate wines have been exported to 26 countries throughout the world. Their only motivation in those early days was to produce good wine for themselves and their friends. Kim had studied engineering and Jeanette was an occupational therapist. They spent three years in London, where Gretchen was born, followed by two years in Madrid, where Karin and Michael were born. When they returned to Auckland to live on the aptly named Vine Street, Kim became a professional photographer. After purchasing the land on Waiheke, weekends were spent on the island building their house and establishing their vineyard and gardens. Everything came over on their yacht, moored in the bay below, including machinery and poles for the vineyard. By 1982, Kim was working full-time on the vineyard, and in 1983 they moved to the island permanently.

Stephen White was the next to establish a vineyard. He had bought his land in 1981 and was commuting from a job in Papatoetoe each weekend to develop Stonyridge. He'd arrive for dinner every Friday night, bearing a bottle of wine, absolute heaven for Kim and Jeanette who, without an income, just couldn't afford the delicious luxury of wine until they produced their own. They remember these evenings with much affection, and learnt a lot from Stephen's youthful optimism and international winemaking experience. Next to plant were Anne and Doug Hamilton of Peninsula Vineyard (now disestablished). Doug was a fabulous problem solver with terrific mechanical skills. Stephen, the Hamiltons and the Goldwaters helped each other out in their new ventures. Kim and Jeanette also received fantastic guidance from

Mark Rattray, a winemaker for Penfolds at the time and from Dr Richard Smart and Allan Clarke, viticulturists who worked for the Ministry of Agriculture and Fisheries.

The Cabernet Sauvignon they planted in 1978 produced a tiny first vintage in 1982. Too tiny, because the local council had sprayed chemicals around their vineyard just prior to harvest, affecting three-quarters of their vines. The Goldwaters took the council to court in a battle lasting six long years – they felt they had no choice but to defend their rights and the future rights for all grape-growers on Waiheke. They won their case and set a tremendously important precedent culminating in the Law of Chemical Trespass.

That first harvest was picked by friends and family – and followed by a delicious home-grown dinner. Word soon spread, and after a few years complete strangers rolled up to their door on picking days. Kim and Jeanette beam incredulously remembering those fantastic days. The pickers feasted on vegetables from the garden and sheep roasting on spits, and Jeanette's 103 plates were used twice over.

They spent the weekend before the release of their '82 and '83 vintages stuffing envelopes containing a brochure marketing their wine. They had a mailing list with over 400 names on it, from people who'd visited them having heard they'd planted grapes. A few days later, a dark blue panelled van approached the house and they were rather fearful of who might emerge from it. But reassuringly, out stepped Roger Price, the presenter of a weekly lifestyle television show, Town and Around. These were the days of just two television channels in New Zealand, so the audience for the programme was huge. Roger explained that they were on Waiheke for a few days and wondered if it would be possible to film Kim and Jeanette. When the programme aired, the first item ended with a close-up shot of the bottle, zooming in on the label. They had posted the brochures the day the programme aired - the timing was miraculous. The two vintages sold out in a couple of days. Gold! But they did have to return many cheques to buyers they couldn't supply.

Internationally well-regarded viticulturist Dr Richard Smart was instrumental in arranging for the International Symposium for Cool Climate Viticulture and Oenology to be held in Auckland in January 1988. It was attended by scientists, producers, leaders and writers from the global stage, including a talented young wine writer named Jancis Robinson. On the day before the symposium began, Smart brought them all over to the Goldwaters' vineyard to relax, meet and greet. They tasted the 1987 vintage and in Kim's words, "it blew everyone's socks off". The Goldwaters were introduced to Geoff Kelly, Bob Campbell and many international wine industry folk. "We had everybody here and suddenly the world was telling us that we were making great wine – they took notice. Our eyes were as wide as they could possibly be," remembers Kim.

That symposium was pivotal for their business, introducing their wines to an influential global line-up of industry leaders. Word-of-mouth marketing grew rapidly and people ventured down their driveway, unannounced, from all over the world. Their first wine export was to Margaret Harvey in London, who was responsible for introducing many Waiheke wines into the British market. One small order was followed by an entire pallet load and suddenly the language of international finance rolled off their tongues – letters of credit and bills of lading. Fascinatingly, a woman named Helen Goldwater was responsible for launching their wine into the American market and, although she's a friend, she is no relation to Kim and Jeanette.

Each year, the couple took their wines to the Wine Expo in Bordeaux and each year, they were invited to dine at Chateau Lafitte, where Jeanette was placed next to Baron Eric. They were visiting most of their 26 export countries on a regular basis. To satisfy the huge market demand, in 1991 they started producing a Marlborough Sauvignon Blanc. They became partners in Rapaura Vintners, with the Babichs and the Spence brothers, whom they describe with warm regard. Board meetings never required a vote from any party – they just kept talking until everyone was happy. Production rose steadily with the acquisition of two hundred acres of vineyards in Marlborough. Goldwater Estate ended up producing 500,000 cases of wine annually, something they never expected – "it just crept up on us".

In the early days, there were four years without an income. But after the successful sellout of their first wines, Kim and Jeanette were finally able to hire one person fulltime and pay their pickers. During vintage, Jeanette became very conscious of the care and attention paid to the work by the woman working next to her. She was Christine Johansen and she remained a central figure in the business for the next 20 years. The Goldwaters' are full of admiration for the brilliant eye and care she took with her work, providing absolute perfection.

The first cellar hand they employed was a local lad by the name of Martin Pickering, whose family have had a house on the island since the 60s. The family had helped harvest grapes at Goldwater's in the early days. Martin was an apprentice dairy farmer, but it was the wine industry that he wanted to commit to. When Kim and Jeanette offered him a job, he leapt at the chance. He is now head wine maker at Stonyridge Vineyard, after many years as head winemaker at Mudbrick Vineyard, and his contribution to wine on Waiheke is enormous.

Work was split between Kim and Jeanette on the vineyard, and most days, they'd only see each other at lunchtime. When Jeanette turned 50, Kim romantically gave her a fork-lift and at 60 she stopped pruning. At 65, bottling line duties were left to the young ones.

Kim always says that in many ways, they entered the wine industry at exactly the right time and they got out at a perfect time. Kim was contacted by an investment company: in a nutshell, their offer was to buy Marlborough and lease Waiheke production, enabling Kim and Jeanette to retain their home on the vineyard. The Goldwaters accepted, and they presumed they would live on the vineyard for the rest of their lives.

But then Kim heard through his daughter Gretchen, a lawyer working at the time for the University of Auckland, that the University was planning to build a wine science programme without a working vineyard. Kim worried about the lack of practical experience and suggested to Randy Weaver, head of the programme, that they use Goldwater facilities. The proposition rose up the official chain until eventually the Vice-Chancellor rang Kim to say they could not accept his offer until it was formal. Kim was out on the tractor mowing a few days later when he had an epiphany that the University should own Goldwater. He knew that the future of our wine industry would benefit immensely from such an asset and so the Goldwaters partially gifted the property. The University now produces wine under the "Goldie Estate" label on site and today the Goldwaters' old home bustles with student life. It was a magnanimous gift – and entirely in keeping with the Goldwaters' generosity throughout their Waiheke residence.

I interviewed Kim and Jeanette in the comfort of their new home, above the wharf at Matiatia. There's room for all the grandchildren and extended family and friends. Kim was off out on the tractor as I left, to mow the large coastal property. Their contribution to our wine industry on Waiheke and internationally is an enormous privilege and a pleasure to witness.

STEPHEN WHITE
STONYRIDGE

–

WINE AND FORESIGHT

In a live television interview in 2014, American Master of Wine Lisa Perotti-Brown asked Stephen White how it felt to be responsible for having changed the perception of red wine produced in New Zealand in terms of quality, style, price and recognition internationally. He was completely blind-sided by the question. He admits to being deeply honoured and taken aback by such praise – but her praise is extremely well-deserved and testament to Stephen's determination and success.

Stephen planted Stonyridge vineyard in 1982 because he had a vision of bringing the culture and lifestyle of the Mediterranean to the Pacific. In the 1970s, after studying horticulture and viticulture at Lincoln University (where an early mentor was Dr David Jackson, a great pioneer of Marlborough and Otago wine growing), Stephen spent three years overseas combining his love of sailing and wine. He joined the Whitbread round-the-world yacht race and skippered boats in the Mediterranean and Caribbean. He spent a winter in Italy working on vineyards, soaking up the lifestyle of olives, grapes and the sea. Stephen began to understand the possibilities of growing all that surrounded him in Italy on home soil in New Zealand. At that time in New Zealand, most of us didn't eat olive oil. Restaurants rarely sold wine, although from 1976 you could BYO – bring your own. Yet the idea of growing olives and vines planted itself firmly in Stephen's imagination.

From Europe, Stephen literally sailed toward his future, working in California en route. His timing was impeccable. In 1976 the legendary Judgement of Paris wine tasting, organised by British wine expert Stephen Spurrier, saw top French wines competing against young new Californian wines, tasted blind by some of the biggest names in the industry from France. A "vinous shot" was "heard around the world", as one observer put it, when the unveiling of the wines revealed the Californian wines, not the French, to be the favourites. It created a huge uproar: new-world wines were at last being taken seriously.

In 1978, aged just 21, Stephen was in charge of the night shift at Santa Ynes Valley Winery in Santa Barbara, California. Late one night he was standing on the top of a very tall tank during pump-over when he slipped and lost control of the pump transporting wine from one tank to another. He battled with the hose while wine sprayed riotously throughout the floodlit open-air winery. He lost 150 litres of wine and spent the rest of the night cleaning up. When the wine makers arrived the next morning, he confessed and was treated to their own woeful tale – while pumping finished wine from one tank to another during a previous vintage, they hadn't realised the pump wasn't connected to the receiving tank and they lost nearly 2000 litres of wine. Stephen learnt a valuable lesson that day which he and his staff still follow. He maintains that successful people say three things more often than unsuccessful people do: "I've made a mistake"; "I don't know" and "I need help".

So with European and Californian experience to call upon, Stephen sailed his way back to New Zealand in search of a hot, dry climate in which to plant the Bordeaux varieties he'd grown to love. His research endorsed Hawkes Bay as the perfect location for his vineyard. But without contacts, family or friends in that area, he felt nervous about starting out far away from fortifying connections. His uncle owned a farm at Cowes Bay on Waiheke Island and Stephen had spent a lot of time there as a young boy, sailing over from Auckland to the island. Visiting his uncle upon his return, Stephen listened to his uncle's complaints about how hot and dry it was for farming, how the cattle dried up earlier on Waiheke and how infertile the soil was. All the wrong reasons for farming on Waiheke were all the right reasons for planting grapes. More research showed him that the island was definitely hotter and drier than Auckland, particularly West Auckland, which had been home to many of our wine pioneers. The allure of growing vines and olives on an island spoke of Italy, while the maritime breezes echoed Bordeaux. Once again, Stephen had arrived 'home'.

With the help of his uncle, Stephen's search for a block of land ended one afternoon up a dusty track on the Onetangi Road. It was perfect - situated on north-facing slopes, protected from the cold winds by the steep flanks of the eponymous stony ridge rising up behind the vineyard. Stephen was one of the rare few back then who knew the importance of terroir. With a loan from his father for a deposit on the land, Stephen purchased the site in 1981.

A few years later, Stephen was still working full-time managing Buckland Kiwifruit Orchard in Pukekohe. He poured money into financing his dream and he needed extra capital. He had a thriving friendship with Kim and Jeanette Goldwater and was their regular dinner guest on Friday nights when he returned to Waiheke to work weekends on his vineyard.

They introduced him to Dr John McCloud, an angel investor and business mentor who loved wine and wanted to become involved. From 1984 to 1988, at a time when Stephen really needed him, John was a wonderful business partner, full of sage advice. Stephen was able to buy John out and, to this day, the business remains 100 per cent his own.

Prior to his first vintage in 1985, Stephen gained further experience in Bordeaux, working at Chateau Angludet and Prieure Lichine in Margaux. There, he immersed himself in the wine style that was to become his life's ambition and achievement. From the very beginning he set out to make world-class, high-quality Bordeaux red wine. Stephen has maintained strong connections to Bordeaux and in 2016 he was admitted into the wine brotherhood Jurade de Saint Emilion.

Stephen and Stonyridge Vineyard implemented techniques and practices that he'd learned overseas, including: planting all of the Bordeaux varieties for blending; selling his wine 'en primeur'; 100 per cent oak barrique maturation; planting the three red Rhone varieties Syrah, Grenache and Mouvedre; planting a commercial olive grove and cork trees; introducing wine tourism to Waiheke; and opening the island's first vineyard restaurant/café.

What attributes, I ask him, do you need to undertake setting up a vineyard and winery? Aside from the obvious qualifications in viticulture and horticulture, Stephen says he gained plenty of the requisite experience by being a professional yacht skipper. First, you have to plan everything around the weather. Then you have to be practical enough to fix a diesel engine or a sewerage problem or whatever other unpredictable, dirty and desperate problems arise on a daily basis. You must also clean up and dress up to host guests including dignitaries and diplomats. Yoga helps to keep Stephen's gaze steady and his manner calm: he is a highly ranked yoga teacher and owner of a studio in Auckland. He's also built a strong core of intelligent support around him, as he nearly lost the business several times, but was saved by brilliant advisors.

The café on the vineyard evolved organically. In the very early days, the Hyatt Hotel in Auckland would helicopter guests over for a vineyard visit; Stephen's then-wife Jane cooked meals to accompany their wines. Wine tourism on Waiheke had begun. Other guests brought their own nibbles with them. Stonyridge's food offerings began with the traditional French jambon – a sandwich with bread, cheese, ham, olives and aioli. Culinary enthusiasts were hired to cater to growing demand. Guests sometimes lost their car keys down the ever-widening dry cracks in the lawn as the grape-growing season flourished. The floors of the open-air restaurant were not swept – they were mown.

A retaining wall was built and a roof erected to keep people dry and shaded. More verandahs were added, and nowadays Stonyridge is open for lunch seven days, private functions, innumerable weddings, outrageous parties and one of New Zealand's most highly prized and hugely attended New Year's Eve dance parties. The venue is all about giving people a great experience – one that's not too formal – "fun with the highest quality" says Stephen. Giving visitors exposure to the ever-widening Stonyridge portfolio of wines is also a key marketing strategy.

In an interview for British television in 1990, when there were just five wineries on Waiheke and only a few more vineyards, Stephen predicted to general amazement that within 20 years there would be 30-40 Waiheke vineyards and plenty of accommodation and restaurants. Our thriving island of wine tourism has indeed become what he'd anticipated.

Stephen's wines have done extremely well and are highly prized by collectors around the world. From the first release, Stonyridge Larose was one of the most expensive wines in New Zealand – and it remains so today. Other Estate grown wines are the Luna Negra - 100% Malbec; The Pilgrim, a blend of Grenache, Syrah and Mourvedre; and The Airfield, a mid-price Bordeaux blend.

Stephen handed over the wine-making reins some years ago. He was burnt out and felt he'd lost the passion required for the job. It was difficult to let go, but he knew that he had to let someone with the requisite fire burning in them take control. Staff remain very aware of the tradition and the reputation they are responsible for. Stephen's current winemaker, Martin Pickering, has deep roots on Waiheke, having worked at Goldwater Estate as assistant winemaker and Mudbrick Vineyard as chief winemaker.

Over the years, when I have travelled up and down the adjacent driveway to my family's vineyard, I've witnessed a never-ending stream of visitors, rambling along in jandals or strutting in the most ill-equipped heels up the driveway to Stonyridge and back down to the bus stop. Helicopters swoop in; celebrities arrive, including ZZ Top, Pamela Anderson and Taylor Swift. Stonyridge is a wine tourism mecca. The man who has created it wryly describes himself as having "always been courageous beyond my ability". But clearly, the abilities and will power of this talented visionary more than keep up with his courage and ambition.

PAUL DUNLEAVY
TE MOTU

—

WINE AND INHERITANCE

When my brother Paul Dunleavy was just a few years old, he ran down the road to the bank in Apia, Samoa, where my family were living at the time. He announced that he wanted to make money and the bemused staff got together and opened an account for him. Clearly, this budding entrepreneur had not yet heard the saying that planting a vineyard involves digging a big hole and pouring all your money into it. Paul ended up working extensively in finance, planting his money in vines and building a future for his family.

Wine has been a part of Paul's life from a young age. Every Sunday in front of his altar boy eyes, a chalice of wine was glorified at church. Our parents had huge parties at home in Napier in the 1960s, and a teenage Paul was fascinated by the role of alcohol in the lively social gatherings. When he was 13, he made his first wine out of mandarins. The following year he acquired a whiskey still, hunted down a recipe, and made a potent brew. My parents were completely unaware of this somewhat dubious hobby that Paul practiced in the darkened country underneath our vast home. Quite possibly the shadows that used to terrify me when I crept under the house as a pre-schooler belonged to Paul shifting the stinky ferment around to avoid suspicion.

We moved to Auckland when our father Terry got a job with Montana Wines. He'd read that children were better able to detect wine faults than adults were, and enlisted Paul to join him tasting wines, and to alert him when he found a fault. Paul also worked school holidays in the warehouse at Montana, sweeping the floors, being Boy Friday. He kept returning to Montana throughout his years at University, where he studied statistics, maths, law and computing. Then Dad was appointed Executive Officer of the Wine Institute of New Zealand in October 1976. Just as I would do many years later, Paul got a lift into University each day with Dad. They had a close relationship, and Dad would discuss his job and the evolution of the industry en route to town. Paul helped Dad in the Institute Office, setting up the computers. He feasted on data, and produced a statistical analysis profiling the production and sales data from the 13 wineries that were members of the Institute. His analysis shed light on the changing preferences of wine drinkers. There was a strong consumer move towards table wine, at a time when the industry had been heavily involved in the production of Port and Sherry.

In 1973, when Paul was 18, Dad gathered together a group of international wine movers and shakers from Europe and America in Auckland. Paul joined the group for breakfast, meeting amongst others, Robert Mondavi, a rising star in the Californian wine industry. Many years later Paul and Robert would meet again, when he helped Paul organise appointments to taste Te Motu wines in California.

In 1983, the government established the Industries Development Commission, which allowed targeted industries, including the wine industry, a very generous rate of depreciation on first-year investments. Paul was working in banking for Broadlands Dominion Group and was appointed the Area Manager in Hawkes Bay, working in farm finance. He approved and processed loans for conversion of farmland into vineyards. While there, he formed a very important relationship with Tom McDonald, a pivotal figure in the national wine industry. Tom had bought his first vineyard in Hawkes Bay in 1927, when he was just 19 years old. He led the way in premium wine production for decades, serving his 1949 Cabernet Sauvignon to Andre Simon in 1965. This venerable French connoisseur of wine and gastronomy tasted Tom's wine alongside Chateau Margaux and was convinced of our ability to produce premium wines. Tom was in his seventies, and he'd have lunch with Paul every week, discussing wine and the industry past, present and future.

Paul decided to establish a vineyard producing quality Bordeaux style wine, which he'd grown to love. He believed that information technology would render many jobs redundant and he wanted to create a business for his future children. He was in the business of entrepreneurship. He observed the clever, brave people whom he approved loans for, in various industries, as they took risks full of confidence in their endeavours. He used them as benchmarks, knowing that, in business, you stake everything upon the skill and knowledge base you've built. And for Paul, that was investment and wine.

In 1984 he was appointed National Lending Manager for Broadbank and returned to Auckland. He began his search for land suitable to plant Bordeaux varietals nearer to his new job than Hawkes Bay. With our brother John (p.30), he already owned land on Waiheke, so it was an obvious choice. Paul's criteria presented just three possible properties on Waiheke, but one in particular was perfect terroir for Bordeaux reds. In

April 1987, he signed the deal for 10 acres on a sub-division of farmland in the Onetangi valley. In addition, he had first rights to buy another 20 acres and in 1988 Paul purchased the entire 30 acres. He talked to Fred Rakich, a notable Chardonnay grower in Kumeu and modelled Te Motu on Fred's closely-planted vines and narrowly spaced rows, unusual at that time. Only grafted vines were planted, but they were in short supply, so Te Motu started grafting their own. My brother John came on board as vineyard manager and together Paul and John learned the ancient craft of grape growing and winemaking.

It took five years to develop the vineyard, planting Cabernet Sauvignon, Merlot, Cabernet Franc and Malbec, plus some Montepulciano and Chardonnay. Paul was in no hurry, preferring to complete the vineyard before embarking on making wine. And he and his wife Avril were still living in Auckland with their growing family, meaning weekends and holidays were spent working on the vineyard. In 1992 they picked a small crop producing one barrel of wine, but the first commercial harvest was in 1993, when the winery was fully functioning.

Paul had a very clear idea that Te Motu would be a négociant as well as a producer. His intention was to make a premium Bordeaux blend, age it for an extended period in barrel, then in bottle and release the wine only when it was beginning to drink well. This method is a financial challenge and takes years to evolve. After 10 years of pumping money into the vineyard, the 1993 Te Motu was released in 1996, with subsequent vintages being held back from sale for increasingly longer periods. An extensive article in Wine Spectator, profiling the rising stars in New Zealand wine, launched Te Motu onto the American market. For the next few years, Te Motu wines were sold to very high-end restaurants and lodges in America, England, France and New Zealand, and ended up on the wine list of the revered Alain Ducasse. Sales were buoyant until the September 11 terrorist attacks in 2001. Restaurant and lodges stopped buying expensive red wines, preferring instead to sell down their well-stocked cellars.

Te Motu had to change its strategy and that meant opening sales channels at the source. In 2002, The Shed restaurant opened on the vineyard for its first summer season. Paul had left his job to plough energy into Te Motu. His days began early with washing and cleaning, hosting tastings and working in the restaurant as maitre'd, enlisting help from his family – his children and his siblings, including me. In-between he'd focus on admin, marketing and sales. He spent time in the restaurant kitchen – his love of fine food and wine is phenomenal, but it was also a financial requisite. Employing great chefs to produce stunning food worthy of the wine has always been a priority and Paul's reward was the pleasure of his guests.

Between 2004 and 2008, New Zealand vineyard production increased from 10,000 hectares to 35,000 hectares. There were more grapes grown than our infrastructure could process. Accordingly, the value of grapes and the land that grew them decreased substantially. That meant great opportunities for investment. Paul had been approached many times by buyers interested in Te Motu, but he had to have a very strong reason to sell. The financial down cycle in Marlborough in 2009 meant that Paul could sell 30 acres of producing vineyard on Waiheke and buy 300 hectares in Marlborough for the same price. The decision to sell Te Motu was a major emotional struggle, but it was financial common sense and would provide great opportunities. John stayed on at Te Motu as vineyard manager and winemaker, so continuity was assured.

Paul bought vineyards in Marlborough while Te Motu was sold to a company that subsequently went into receivership. A few years later, Paul and family, together with a group of investors with a passion for Te Motu, bought the vineyard back. With fresh ideas and energy, Te Motu and The Shed are again operating with their original personality and philosophy. Paul's dream of creating a wine business for his family is flourishing in reality. He has plans for further investment on Waiheke and in Marlborough, fully aware that to be financially sustainable in the wine business you need to secure and produce adequate volume.

In the circle of life, Paul has a seat on the Board of Directors for New Zealand Winegrowers, just as his old mentor Tom McDonald had done. After 44 years, he's still helping Dad organise national wine competitions, and they're always tucked away, deep in discussion about the fascinating industry they've devoted their lives to. Paul has embraced his inheritance from Dad and maybe one day his four children will do the same.

JOHN DUNLEAVY
TE MOTU

—

WINE AND INVENTION

An accident changed the life of my brother John Dunleavy. He was working as a builder, having changed career from teaching art, when he nearly sliced three fingers off using a power saw. While he was in hospital, our brother Paul (p.26) asked John to quit building and manage the block of land he and others had bought in the Onetangi valley, to establish a vineyard, which would become Te Motu Vineyard. With no knowledge of grape growing or winemaking, John and his wife Deborah came to Waiheke in January 1987 with their young son Sam – and they're still here three decades later.

John had endless practical capabilities and advisers to help him in his new endeavour. Our father Terry Dunleavy was, at that time, the CEO of the Wine Institute of New Zealand – now Winegrowers NZ Inc. Paul had worked in the wine industry during university holidays and had always dreamt of making wine. They also brought in consultant Allan Clarke to help set up the vineyard. A local contractor started laying out the posts, while another ripped the soil down the rows so that the vine roots were more easily able to penetrate into the clay. The cast was finally cut off John's arm, Debbie gave birth to their 2nd son Tomas, and their lives began anew.

Te Motu is planted in the classic Bordeaux mix of Cabernet Sauvignon, Merlot, Cabernet Franc, Malbec and the Rhone varietal Syrah. Initially they also had a small amount of Chardonnay planted, which they swapped for the finished wine with other producers on the island. Te Motu translates as 'The Island' acknowledging the original name for Waiheke: Te Motu-arai-roa – the long sheltering island.

The first commercial harvest was in 1993 – it had been a great growing season, but preparations for vintage didn't go to plan. Terry had bought an old French destemmer/crusher that arrived a few weeks before vintage. But it was in unexpectedly bad condition and was sent back to town to be sand-blasted, spray painted and undergo full bearing replacement. Then the unthinkable occurred and it was dropped on its way, splitting the casing and buckling the drive axles, bearings and housings.

Vintage was days away, and their critical piece of equipment would take up to ten days to repair. With no other options available, tensions began to rise. The machine did arrive on time – but would it work? Family and friends waited patiently for the inaugural harvest to begin. Finally the truck arrived to move the heavy machine into position. The electrician connected it, they plugged it in, took a deep breath, pressed the button... and away it went. John says his curly hair changed to grey that stressful week prior to harvest. Some years later, that 1993 Te Motu was awarded a High Silver Medal at the International Wine Challenge competition in London – a very prestigious accolade, followed by Gold Medals for the 1994 and 1996 vintages. John had learned fast.

In 1996, Te Motu was visited by Dr Paul Pontallier, chief winemaker, Wine Director at Chateau Margaux in Bordeaux, until his untimely passing in 2016. Chateau Margaux, established in the 17th century, is one of the famous First Growth classified wines of Bordeaux. Every year they invite a few guest winemakers to work with them through harvest and vintage – and that year, Dr Pontallier was so impressed with Te Motu wines he invited John to join them in Margaux for vintage. John arrived in Bordeaux on a Sunday afternoon in September. Anyone who has travelled to France will know the quietude of a Sunday afternoon. Shutters are closed, streets are empty and you wonder if you're the only person left on the planet. Unbeknown to visitors, on unseen patios and in gardens, long lunches and hearty conversations are taking place. John alighted from the train not sure of where to go, and he says he was eyeing up a nice little spot between the vines to set up for the night before he found someone to point him in the right direction. He entered the Margaux gates only to be greeted by an Australian accent yelling from across the yard: "you must be the Kiwi – you're staying over here with us".

For the next 6 weeks they worked day and night. John discovered that the way they made their wine in Margaux was the same way he made Te Motu wine – an enduring assurance. He was also impressed by Dr Pontallier's respect for the human touch in winemaking, honouring long-held traditions over mechanised efficiency. The owner, Corrine Mentzelopoulos, would point at John's bare legs with amusement as she passed by, as no one else wore shorts. One of the delights when participating in harvest in France are the long lunches and, at Margaux, a five course meal is served. The owners sit with a dazzling array of guests – politicians including the Prime Minister of France and wine writers. The guest winemakers sit at the next table, and are handed the remains of the many legendary wines opened for the esteemed guests.

Another highlight for John was working with the cooper or barrel maker, who had inherited the job from his father. He and John couldn't speak each other's language, but even so, John helped him to repair barrels, using much admired specialist tools inherited over generations. John is an incredibly gifted carpenter and designer, crafting beautiful tables out of old wine barrels. He builds benches, houses and workshops, and transforms interiors brilliantly. He designs and casts basins and tables from concrete, and creates gorgeous lacquered flooring. He designs Te Motu wine labels and is a wonderful gardener and landscape architect.

John dreamed up many practical solutions for use on the vineyard. He was watching birds pecking at the fruit behind the nets draped over each individual vine and wondered how he could minimise the damage. He imagined joining two nets together and draping them over three vines to create a wider protected canopy area where the nets were not sitting on the fruit, while also eliminating the need for one net per row. Larger nets were not available in New Zealand at the time, so the nets were stitched together.

He also designed an over-the-row sprayer back in the days when the Clean Air Act came into law, after complaints about spray drift. He made a shield that hung on either side of the vine and had tanks in the bottom. The spray hits the internal walls of the shield and is pumped back into the tank, saving 60% of volume previously used as well as eliminating spray drift. It received quite a bit of interest, but what he really needed was someone to come along and give him capital to develop it to the next stage. Others beat him to it and developed the concept and nowadays it's an everyday piece of vineyard equipment. He also designed a tractor attachment to dig out young vine rootlings in the nursery, eliminating the need for hand digging.

Not only does he take care of the vineyard and winemaking, he also built the restaurant and everything in it. The Shed literally was the tool and tractor shed, where we'd escape the weather at lunchtime. It was home to the Friday gatherings of vineyard folk on the island, for an end-of-the-week collegial wind-down and many harvest parties. When the time came to open Te Motu to the public, John designed and rebuilt the existing shed to become a modest restaurant, spilling out into the garden. It is now a lauded food and wine destination.

John is the longest serving day-to-day vineyard manager/winemaker on Waiheke and has been mentor to many. He and Matt Allan, who has also been a vineyard manager/viticulturist at Man O'War for over 20 years, have learnt a lot from each other. Likewise, Herb Friedli was a great early mentor for John – very supportive and immensely knowledgeable. John has also learnt from talking to the vastly experienced Danny Schuster, consultant to the famous Ornellaia in Italy and Stag's Leap in Napa, who has made Waiheke his home in between jobs.

John can simply look out at the day and gauge what weather lies ahead. He's monitored significant changes in the weather patterns over the years. In the early days, the only trees on the property were casuarinas, leaning over from years of being blasted by the predominant sou-westerly winds, which blew for weeks on end.

Nowadays, the casuarinas are growing evenly on both sides and stand straight: the winds have evened out. Rain patterns have changed – the wet season used to start in June and continue for months. Some days you couldn't drive up to the vineyard because the road was flooded. Nowadays the winters are arriving later and overall it's drier and warmer all year round than it used to be.

John and Deb's son Sam has completed a post graduate Diploma in Winemaking and Viticulture and is working at Te Motu - a third generation taking wine into his future. John has watched the industry grow and welcomes the influx of younger professionals, with training, knowledge and international experience.

John used to pull at that grey hair of his, but is now more philosophical, knowing that if he has done everything in his power to ensure excellence, his job is done on this land he knows so intimately. He is well past feeling responsible for nature wreaking havoc, accepting that it can destroy your entire year's work in minutes, but that you cannot control it. Walking out into the vineyard on a beautiful day is all it takes to feel a contentment that eases the years of stress and strain. Every good vintage is a highlight, and opening another great bottle of hard work and dedication is an absolute joy. He is a quiet and humble guy, who will find it hard to read about himself and acknowledge that his tenure, creativity, friendship and experience are celebrated and honoured by many.

BARRY AND MEG FENTON

—

BARRY AND MEG FENTON
FENTON

—

DESTINATION WINE

Barry and Meg Fenton have made bold and creative decisions all their lives. In 1969, they and their two business partners opened a new travel company offering package tours. Within six months they had doubled their annual turnover. The years rolled on and they franchised the business until, from four people, they grew to 300 throughout New Zealand, all working under the banner of United Travel. Another decision was particularly fabulously bold and slightly scary: Barry and Meg bought a large headland property on Waiheke Island just after the global share market crash in 1987.

Meg had been working in the travel industry in Fiji in the 60s. On arrival back in Auckland in the late 60s, she met her boss, Mr Barry Fenton – and he didn't stay "Mr Fenton" for long. The couple were keen sailors and spent weekends and holidays sailing around the Hauraki Gulf. These were also the days when everyone usually travelled by long sea voyages to their chosen destination. London was four to six weeks away. The pleasure of your holiday was the journey itself – it had to be. But the young couple could see the future of travel changing: the jet plane was opening up an abundance and profusion of choice. Barry and Meg scouted the globe to find the most alluring destinations – once distant, now within reach. Customers flocked.

Once their travel agency took off, international holidays for these professional travellers involved staying put in one location only. It was luxury not to have to follow a rigorous timetable of looking after clients in city after city. In the early 70s, they enjoyed the islands of Bali long before other tourists. Later, they'd make Maui their island of relaxation in the midst of a very busy year. It seemed that island life suited them perfectly.

In 1975 their son Matt was born. As he grew up, it turned out that he wasn't a keen sailor at all and Meg suggested they buy a holiday home by the sea with easy boat access. They looked at possibilities in Coromandel and the Bay of Islands, but the Fentons had long had their eye on Waiheke. Barry and Meg had visited the island in the early days of their relationship, camping at the southern end of Onetangi Beach. In the early 70s, Barry and his business partner Ollie Newbegin actually bought land on Waiheke Island. With great foresight, they had plans to develop a golf course and a hotel in Onetangi. Investors nowadays will scratch their heads to read that, back then, the pair weren't able to raise the finances to develop the site, due to the limitations of transport to Waiheke. The travel agents were told there wasn't enough travel available. The land was eventually sold and is actually now home to the Waiheke Golf Course.

They were keen to buy on Waiheke again. Their business partner had bought a headland property on Waiheke and they could see the potential. Meg read an advertisement in a Saturday paper for a north-facing headland property on the island that was being auctioned the following week. On the Tuesday afternoon, they picked Matt up from school and ferried over to have a look. It was much more than they were looking for in every sense and they were completely bowled over by it. The auction was tipped to have a lot of interest from far and wide, but they stood alone, the only bidders in the room. The world share market had crashed. Blinking with disbelief they became the proud new owners, naming their price for a magical property, close to the ferry. The first thing they had to manage on their new property was the felling of five acres of mature pine trees. The land was exhausted by the pine and lay fallow for a few years.

By then, Barry and Meg had begun to enjoy the very few New Zealand Bordeaux blends on offer, heading south down the motorway in Auckland to visit deRedcliffe Estate, tasting the wines made by Chris Canning. When Chris opened Hotel du Vin on his property in Mangatawhiri, they booked in for business conferences. On Waiheke, they drank Goldwater and Stonyridge whose wines were raising the profile of quality Bordeaux-style reds. They got to know Stephen White, whose 1987 Stonyridge Larose had impressed serious wine palates internationally, earning him critical acclaim. After discussions with Stephen about their availability of land and his desire for more fruit, they decided to pull together. Stephen would take care of all the planting, viticulture and winemaking on the Fentons' land and they would split the wine 50/50. They planted Cabernet Sauvignon, Merlot, Malbec, Cabernet Franc, Petit Verdot and the inaugural vintage of Fenton Estate was produced in 1993.

Barry is a smart guy – he knew that his winemaker's profile was key to wine sales. Stephen White was a remarkable guy to have made their wine in those first three years, earning their 1994 Fenton gold at the Air NZ Wine Competition. And Stephen was followed by another significant name in New Zealand's winemaking history: Kim Crawford. It's not that it was hard to sell quality red wine in New Zealand in the 90s. Demand was good, and supply was often by allocation from very small producers. But that wasn't a reason not to aim high, and after the Fentons doubled the size of the vineyard to nine acres by the mid 90s, there was a lot more wine to move.

An important reason for the Fentons to enter into the wine industry was to introduce a different circle of friends into their lives. They'd been all-consumed by their travel business, with many close friends connected to it, and wine was a great opportunity to branch out. Barry would invite people over to their Waiheke home on a Friday afternoon, each bringing a bottle of wine shrouded in a paper bag, to be tasted and discovered – each trying to outdo the other with many laughs and lessons in-between. He got to know Bob Campbell MW and sometimes Bob would join the group, as would Stephen White and Kim Goldwater. Barry recalls valuable guidance from Bob, such as how to compare customer and producer preferences in wine styles. He also introduced Barry and Meg to John Hancock from Trinity Hill, who was to become their next winemaker.

In 1991, Barry got a phone call one Monday evening. Meg could see that he was in a bit of shock. It was his daughter Jane and she was 24 years old. They knew about Jane, who had been adopted out at birth, before Meg and Barry had met, but had always respected the wishes of her adoptive parents. Meg and Barry had often mentioned her, wondered about her, but knew they could not intrude. After her adoptive father passed away, Jane decided to make contact and a few days later she came to visit them on Waiheke Island. It was an emotional roller coaster for them all. They'd never told Matt about her, preferring to wait until he was old enough to understand. When they did, this teenager's response was phenomenal – he hoped his Dad had done right by her. He listened to the circumstances and understood. When Jane described him as her half brother, Matt wasn't interested in having a half sister – he just wanted a sister. Friends dropped by in the middle of lunch unexpectedly, overhearing Matt on the phone to his friend, telling him that he couldn't talk because his sister was there. Looks were traded between their friends – what was going on? They were soon enlightened and Jane stepped into the family and has been part of it now for more years than not. Both Matt and Jane have had special edition Fenton wine labels named after them and their children.

Barry and Meg are completely down-to-earth people, warm and engaging. They enjoy drinking red wine and bought the Bordeaux First Growths back when they were much more affordable than they are today. Irreverently perhaps, Barry remembers, with fondness, getting together with his neighbour, Robert Gilmore, for brunch. They'd make a dish of black pudding and bacon, matching it with a bottle of the illustrious Petrus. He did it because it was delicious and because, outrageously, they could. He follows the European tradition and is adamant that red wines should always be consumed with food.

After John Hancock went on to the next stage in his stellar wine career, Barry approached David Evans-Gander from Passage Rock Wines on Waiheke. Meg and Barry were winding down their business obligations and wanted to divest themselves from the responsibility of the vineyard. They needed someone to take it on for the long-term and David was the perfect choice. Passage Rock wines were claiming medals and they were keen to grow their production. A deal was signed enabling Passage Rock to take over complete management of the vineyard in return for wine, sold under the Fenton label. They've top-grafted some of the original Bordeaux varietals with Syrah and added some Montepulciano to create a blend especially for Barry.

So the couple are happily retired, busy as ever, spending quality time with family. Barry began painting many years ago, moving onto sculpture, and their Waiheke home is testament to his talents and his great sense of humour. Ever seen a pink Cadillac convertible with a dozen long legs popping up from the back seat? You might see Barry driving around in it. They're deeply engaged with cultural philanthropy on Waiheke, especially in visual arts. Barry and Meg are keenly aware of giving back to this inspirational and creative community, and their commitment helps us to continue to prosper.

ROBYN JONES

MUDBRICK

—

WINE AND ROMANCE

In 1991, in partnership with her then-husband Nick (p.46), Robyn stood on a bare headland in Church Bay and dreamt of spending her days tending vines and gardens and raising a family. She didn't envisage owning a restaurant, let alone one of the most popular vineyard restaurants and wedding venues on Waiheke. But in order to survive, she had no choice but to run with it. There were times when she desperately wanted to walk away; other times when she's wanted to be nowhere else. In short, it's been a roller-coaster ride creating Mudbrick restaurant, where Robyn enhances each special occasion with her love of romance.

Robyn's father was a vet in Bombay, south of Auckland, and Robyn grew up with chickens and horses on a big block of land. Her father also operated dog-boarding kennels and Stephen White lodged his dog with them. He'd stay for dinner from time to time, telling them about his vineyard on the island they all adored – Robyn's family had a bach in Onetangi on Waiheke. Later on, Robyn's mother lived on the island full-time, and on her suggestion, Robyn and Nick visited Stephen to discuss their growing fascination with wine.

Robyn and Nick were living in Auckland – he worked as an accountant and she was studying to become one. She'd worked in beauty therapy, but wanted more depth from her profession. "It was a big deal for me to attend university - I never imagined I would," she says. But after completing her degree she realised accountancy was not her future. Over dinner one night she suggested to Nick that they buy a block of land on Waiheke and plant a vineyard. Her father had recently died and had left her enough money to invest in land.

She pauses amidst our conversation, startled at how crazy that idea seemed at the time. But Nick wasn't enjoying his job either and was open to the notion of starting their own business. Keen to make that change quickly, Robyn began looking for land, eventually discovering the Croll Farm sub-division in Church Bay. She did what she freely admits nobody should ever do when buying land to plant grapes – she fell in love with the view. They considered the cold westerly winds that sweep across the property, but resolved to build shelterbelts, planting hedges to overcome the weather. They stand tall and proud to this day, providing the necessary protection.

Robyn was looking forward to spending more time with her mum once they'd moved to the island but sadly, her mum died in 1993, two years after Robyn's father. Her death instilled in Robyn a yearning for family, and the weekend Robyn and Nick moved to Waiheke in October 1994, Robyn discovered she was pregnant. The vines were planted, they were living in their newly built Mudbrick home, Nick continued commuting into town for work – everything was going to plan.

Then it got tough. Dr Richard Smart had visited many vineyards on the island to provide them with an analysis of their businesses. He gathered everyone together one fateful day and declared that, with their small vineyards and distance from Auckland, unless they added a hospitality side to their business, they would all fail. Stonyridge was the only vineyard restaurant operating on the island in 1994. But Richard asked everybody basic questions - where they would sell their product; how they would market it? Did they imagine, he asked, that people would flock over to Waiheke to buy their product? Robyn was hugely pregnant and found Richard's honesty deeply depressing but, in hindsight, she sees it was extremely constructive. It annihilated all of their fantasies – "these fledgling vineyard owners running on dreams". It was 1994 and they had big, bold decisions to make.

Robyn is not an extrovert character who commands the restaurant floor. She prefers socialising at home and spending time in her garden. So the idea of starting up a restaurant in the house she imagined herself and her family living in seemed outrageous. But the choice was clear – they would have to lease out the vineyard and go back to town to live and work, or start up a restaurant. She tells me that she has always been struck by self-doubt, which seems hard to believe in the light of day now. Many years later, upon reflection, she knows that "if you have the smallest feeling that your idea will work out you should go for it. You just need to surround yourself with the right people to make it happen". In 1995 after borrowing money, the restaurant opened – and from the beginning it was a success. She had lots of ideas that Nick was brave enough to agree with and helped bring to life.

New Zealand had won yachting's America's Cup in San Diego. Our defence of the cup brought a global village of yachting teams to live in Auckland and prepare for the next cup challenge. They visited Waiheke often. The recession faded, people wanted a destination and Fullers Ferry

Company wanted reasons to bring them over with their new fast ferries. Historically, Waiheke was changing forever.

Mudbrick's first vintage was a year later, in 1996. Nick and Robyn had planted Cabernet Sauvignon, Merlot, Cabernet Franc and Malbec. They took their grapes to Goldwater Estate, who initially made their wines. Nick looked into the tank after crushing the first small pick of Malbec wondering where the juice was: it was sitting in the lines connecting the crusher to the tanks. Robyn roars with laughter now at how naïve they were. The next vintage wasn't much bigger than the first. When tasting the wine with Kim Goldwater they asked whether it was as good as his wine and Robyn remembers Kim rolling over laughing.

In 1996 Robyn gave birth to the couple's second child, and the restaurant was seasonally busy. I know – I was working there, hosting wine tours and tastings and working in the restaurant. Customers would arrive, fresh from the ferry all at the same time, needing to leave all at the same time. This was before Eftpos and we were armed only with the old zip-zap machines to process payments and get everybody out in time for the ferry. Entire evening wedding parties would arrive before the departure of an earlier wedding or the lunch service. It seems hilarious now, but add in continual power cuts and missing deliveries for the kitchen that were inevitable for Waiheke back then and, gosh, it was stressful!

Robyn had to find out the hard way how to delegate. She was managing the office with two young children, in a growing business and began to feel guilty about having a nanny. They were unheard of on the island at that time and Robyn felt she was being judged for it. She wanted to spend more time with the kids, but agonised over leaving the office, because no matter how many hours she spent in there, it would never be enough.

Then one day, a virus attacked their computers, wreaking havoc. Frustration and anxiety combined with parenting guilt came crashing down. She arrived home from work, said good-bye to the nanny, lay down on the floor and dissolved in tears, with her kids climbing over her asking "mummy what's the matter". From that day on, she changed her priorities: she would work mornings, hire someone for the office and spend afternoons at home with the kids. She was learning to delegate.

Mudbrick was her other child. It took ten years to grow up and she is amazed at how things started to flow more easily after that decade. She never imagined hosting her favourite celebrities – the Coronation Street stars and in recent years, Cindy Crawford and Taylor Swift. Mudbrick has hosted many first dates and proposals, and is a renowned venue for weddings – at its height, 100 weddings annually. Robyn's satisfaction comes from creating an environment steeped in romance, for guests to enjoy while celebrating. She has a great sense of style, evident throughout the restaurant and ornately-manicured gardens with extensive topiary. She's attentive to every detail from taps, door handles and lighting, to leather couches draped with blankets in front of a roaring fire.

She never had to look after the money side of the business until she and Nick separated in 2009 – but actually, she has enjoyed using that accountancy degree earned all those years ago. Robyn describes how the couple met when she was 19, "just a child really". Much later they realised they hadn't lived their own lives and wanted to explore their own paths separately and see what that might reveal. They continued working alongside each other – overcoming plenty of ups and downs while learning to accept the changes in each other and to acknowledge the person they'd separated from. When I interviewed them, they were in a huge phase of productivity, building accommodation and the new Archive Bar in the exquisite gardens looking out across the Gulf. They are moving forward side by side.

Bravery always pays off says Robyn – she tells that to her kids. She is proud of everything they've accomplished. She embraces her growth in confidence alongside the business and upon reflection is impressed by how much time has changed that young woman discovering a bare block of land. She nods toward her younger self, struggling with self-doubt, asking herself: "who is that, was that you"? This quietly contented achiever has plenty of ideas to keep her fulfilled in the business of creating romance and enduring memories for decades to come.

NICK JONES

—

NICK JONES
MUDBRICK

—

WINE EQUATIONS

Back in 1991, Nick Jones decided he was not happy about being an accountant for the rest of his life. So he got out of that office and onto the land, launching into exciting and terrifying territory where the crunching of numbers remained an awfully necessary skill.

In 1988, he and Robyn (p.40) toured Europe by campervan, and tasted the wines of France, Spain and Portugal. After their return home, Nick enrolled in an evening course that introduced him to wines from around the globe. He began to build a small cellar at home, with an emphasis on buying wines he'd never heard of in order to continue broadening his knowledge.

When he and Robyn started visiting her mum in Onetangi, he walked slap bang into his new interest – our burgeoning wine community. He met Doug Hamilton from Peninsula (no longer in operation). He met Stephen White, Kim and Jeanette Goldwater and my brothers – John and Paul Dunleavy. Robyn suggested they move to Waiheke, plant grapes and make wine. Their villa in Auckland was worth the same amount as the Church Bay property she had found – 25 acres of prime real estate, which, in the midst of an economic recession, was an audacious and brilliant buy. Spotting a business opportunity is something that Nick is really good at, and the couple's timing was perfect. In 1993 they planted the first vines and moved into their Mudbrick home, which ultimately became Mudbrick Restaurant.

By 1996 Nick had quit his accounting job to manage the operations of both the vineyard and restaurant. He remembers it as being an exciting, scary time. They had so much to learn. With horror they observed the tiny batch of Malbec being pressed in their first vintage. Five years on from buying the land and planting the vines, they were the proud owners of five barrels of wine – just 1500 bottles. It was the restaurant that ensured survival. At the time, Mudbrick was the only option for fine dining in a vineyard restaurant close to the ferry. And equity in their land had grown, enabling them to reinvest capital into their business.

Growing numbers of visitors wanted to see the vineyards and the Fullers Wine Tour was born. The daily tour stopped at three vineyards, ending with a visit to Mudbrick, introducing visitors to their spectacularly-sited restaurant and vineyard. Initially they poured Goldwater wines for tasting until they produced enough wine of their own. Nick remembers how interested and intrigued the visitors were in this young couple and their endeavours.

As Nick reflected on his past, a sort of shocked surprise flashed across his face when he considered some of the decisions he'd made. Just one year after producing only five barrels of wine, he decided their yields in 1997 would be large enough to justify building a brand new winery and hiring a full-time winemaker. Harvest came in producing just seven barrels of wine, which were lined up neatly in a small corner of the long echoing chamber.

While Nick and Robyn sought advice, not all that advice was successful. After 20 years of struggling with a south-facing block planted ill-advisedly with Cabernet Sauvignon, it was finally replanted with Pinot Gris, more suited to the cooler site. Nick acknowledges that it took them eight or nine vintages before they really understood how to grow the grapes, when to harvest, how to prevent disease and finally, how to make wines that reflect what you are trying to achieve. And they weren't alone: many were new to the industry. Plus, the tried-and-true methods of winemakers and viticulturists newly arrived from Hawkes Bay or Marlborough didn't necessarily suit Waiheke grape growing conditions.

Nick remembers another brutal experience – a blind tasting where he'd entered his first Syrah from the disastrous 2001 vintage. Upon tasting it, Kim Goldwater announced firmly that if he had produced such a wine he would go ahead and shoot himself. Nick was mortified. He's able to share that memory now because, for many years, their Syrah has consistently been selected by Bob Campbell to be in the First 11 New Zealand Syrahs team in the annual Tri-Nations competition against Australia and South Africa. Deservedly, Nick feels a deep sense of pride in how far they've progressed. Many Mudbrick wines have received outstanding recognition.

Nick had also been observing his friends' vineyard in the Onetangi valley – now Wild Estate. They had a sheltered site and their first vintage had yielded 15 barrels of wine. So in 1998 Mudbrick purchased Shephard's Point next door to Wild Estate –12 acres now planted in Chardonnay, Viognier, Cabernet Sauvignon, Merlot, Malbec and Syrah. They built a winery on site, which they lease out to a few other vineyards for shared use. They also have a small vineyard planted with Tempranillo and Chardonnay on their neighbour's block in Church Bay.

The better known you become, the better the quality of people that you attract. They've had three significant winemakers: James Rowan, Martin Pickering and currently Patrick Newton, all of whom remain great friends with Nick. "They are all very successful and confident winemakers with superb palates - great guys with very different personalities", says Nick. James went on to become a judge for the Air NZ Wine Competitions and consultant winemaker to many vineyards, including Soho's Onetangi vineyard. Martin is the now chief winemaker for Stonyridge Vineyard. Patrick was named New Zealand Young Winemaker of the Year in both 2014 and 2015.

Nick used to feel like the new kid on Waiheke, but 25 years down the track he feels like the elder statesman. He's really optimistic about our wine industry and feels that it's matured and flourished and is glad to see the years of struggling, learning and adapting fade into the past. He's discovered he's good at listening to other people seeking his advice and enjoys helping them to make the right decisions. He's a firm believer that his greatest asset has been the land. It has provided equity and cash flow, enabling him and Robyn to enjoy success. Although they are no longer married, their business partnership is thriving.

He ends our interview by saying that wine regions are great little communities. Does he think it's still possible for new people to come in and start from scratch – "yep, absolutely". Is it a good idea to get Nick to check out your business plan – "yep, absolutely". Will he will be there to share his hard-earned knowledge with you – "yep, absolutely".

MATT ALLAN

—

MATT ALLAN
MAN O' WAR

—

WINE FOR LIFE

At the tender age of 17, Matt Allan had to make a choice that would change the course of his life. He already knew horticulture was for him – he had left school at 15 to complete a one-year horticulture programme at Gisborne Polytech. And, as top student, he had landed a prize job working on a mixed orchard with a small vineyard. But the property was being sold to Montana Wines – now Brancott.

The question was: should he stay on with Montana Wines or leave to continue working with sub-tropical fruits?

His parents had a lifestyle block, planted in avocados and citrus trees, but Matt remembers his dad saying that wine had been around a long time – that he couldn't go wrong with it. So Matt chose to stay, and Montana kept the vines, ripped out all the fruit trees, punched down posts and planted more vines. Matt learnt everything from the ground up under the wise mentoring of Warwick Bruce. During harvest he'd accompany Warwick on his quality control of contract grape growers' vineyards, assessing the fruit, sampling juices, talking to the winemakers like Steve Voysey.

After five years he wanted to step up, and there were no management vacancies on the horizon at Montana. So he answered an advertisement in the local paper for a job establishing five acre of vines, with room to grow, on some random island he'd never heard of called Waiheke. Thus in 1993, at age 22, he found himself alighting the ferry at Matiatia Wharf on possibly the worst day of weather Waiheke has ever seen. And Matt should know – he's been out in our weather for decades. He was met by Kim Goldwater who was consulting for Matt's new employer, the Spencer family, owners of a 4500-acre farm at what we locals call "the bottom end" of the island. They were looking to diversify into grapes from cattle farming.

As they drove across the island in Kim's tiny yellow van, rattling along gravel roads and winding their way around the rolling hills, Matt could not see himself planting grapes anywhere. His heart was sinking fast. But as soon as they arrived on site his spirits lifted. They drove to where his future home stood at Man O' War Bay. (The bay was named by Captain Cook in 1769, who noted in his journal that the kauri trees he saw on the cliffs would be ideal wood for the masts of the Man O' War battle ships of the Royal Navy.) Then Kim and Matt carried on inland to a sheltered site where people were busy banging posts in the foothills of Stonybatter. And that initial vineyard still exists today. Matt was very excited by what he saw and immediately felt a connection to the land. Waiheke became home.

At that time there was a farm manager, a shepherd and Matt. When there was downtime on the vineyard, Matt helped out on the farm with the fencing and the cattle. He enjoyed the work diversity and acquiring new skills. He was also happy with his own company, but when in need of a sympathetic ear, he'd visit Jeanette Goldwater who'd sit him down for a cup of tea or a glass of wine and a home-cooked meal. She was a life-raft for him, and to this day Matt is convinced that she has no idea how important her presence was – and still is – in many peoples' lives.

Matt initially planted Cabernet Sauvignon, Merlot and Cabernet Franc, and then planted another couple of acres in Chardonnay. There were only a few Waiheke Chardonnays at that time including Goldwater and Delamore. The team expanded to include Luc Desbonnet as winemaker. He had just returned from studying and working in the wine industry in Central and Eastern Europe, bringing on board a wealth of international experience.

Then, Berridge Spencer returned to New Zealand after studying in the USA, committed to grow the operation to the next level. Expansion began – not by two or three acres at a time but excitingly by twenty or thirty. Pinot Gris was planted in the early 2000s – not on Waiheke but on one of its satellite islands, Ponui, of which the Spencer family own one third. Despite the entire off-shore operation being difficult and very costly, the site was chosen because of its viticultural potential. In 2005, they planted another inland Ponui block with Cabernet Franc, Merlot, Malbec and Syrah, so hidden and protected you'd never know it existed.

The name Man O' War became available for use in 2005. The timing was perfect: the newer wines were coming on stream – wines which had benefitted from years of research into where and how the vines would be planted, with closer planting of rows and tighter density of vines. Advanced cultivation of root stocks and clonal selection produced much better outcomes in the vineyard and continual improvements in the wine. A series of labels were created, connected to the nautical theme: Dreadnought; Ironclad; Warspite; Pinque.

The Exiled Pinot Gris is made from the St Helena clone, named for the island where Napoleon ended his days.

What makes Man O' War unique – not only on Waiheke but in New Zealand – is its size and scale over diverse terrain. It includes over 75 single vineyard sites – an enormous number, some clinging to hillsides above the panoramic coastline. The winery is crazy busy. Each different vineyard, including some sub-plots and different clones for all of the red varieties and the Chardonnay are fermented separately. They have up to 100 ferments ranging in size from a single barrel, to four tonne, to even larger quantities. In 2008, Duncan McTavish joined the team as winemaker. Matt describes him as a great craftsman trying to convey the purest expression from each vineyard. The winemaker's skill lies in blending the right parcels of finished wine together, to create the levels of quality desired. The benefit for the team is that they can taste each different site and begin to understand the potential reflected.

Just recognising where all the different vineyards are and what they're named takes at least six months. There are nine fulltime employees and finding long-term additional workers was no easy task. But in 2005, via a labour contractor, the company invited a group of Ethiopian refugees onsite to see if they might like to work and live at the vineyard during busy times. They were a great fit, and return year on year. Around 25 join the fulltime workers for harvest, pruning and summer canopy management for about 7 months each year. They have a full-time cook and pastoral carer. Their foreman has liaised on behalf of the group since they started travelling to Waiheke, and Man O' War's investment in him and his team has been rewarded with loyalty.

The difficulties of navigating the dangerous slopes require tailor-made German tractors. On this vineyard you cannot just send someone out to do a certain task, knowing that they'll be safe. Matt and his team carefully assess and improve the level of experience and confidence of each worker. An additional layer of management is required to assess each decision about who goes where to do what and when. It's tough yet rewarding work – some leave, some leave and return, and some stay on for a decade or more.

The Tasting Room started life as a portable caravan in 2006, sitting alone on the flat spread of green lawn a few metres from the shoreline of Man O' War Bay. It was a feasibility study, as they didn't think people would venture the distance to taste and buy wines. They certainly didn't count on the amount of boating traffic. Word quickly spread as more and more of the maritime community dropped anchor and popped ashore to replenish their wine supplies. One caravan became two and then a permanent tasting room, providing light meals. A chapel has been built in the same colonial style as the old family home and the Tasting Room and when you're relaxing outside, you'd swear it has been standing quietly for generations, a remnant from the early settler days.

In 2010 Man O' War wines launched internationally. Duncan introduced the wines in Asia while Matt did the same in America. They met in London to showcase their wines to the trade, including a memorable evening at The Providores Restaurant, owned by renowned New Zealand chef Peter Gordon. Matt was never comfortable with public speaking. He'd look out into the audience gathered, recognizing big names in the trade. He'd wonder why they'd want to listen to him and what he could possibly tell them that they didn't already know. Until, that is, Berridge Spencer gave him the best piece of advice, which was simply to tell the Man O' War story – from five to 150 acres planted with small vineyards, spread across 4500 acres of land, on two islands. It's an inspirational story, and one that's been lived by Matt day in and day out from the outset in 1993.

Matt, like so many others in our wine community, is a humble guy. He lives with his wife Denise whom he met on the second tee of the Waiheke Golf Course when he was 25. They have two daughters, Brooklyn and Gracie. He's part of a small community of families who work at Man O' War and in the larger wine community. Wine has been his life now for 29 years and he certainly can't see himself being anywhere or doing anything else. Being intimately involved with the entire process is indelibly satisfying. Producing wines good enough to stand on the international stage is what gets him out of bed in the morning.

Matt's had tremendous support from the Spencer family, acknowledging he's a better person both professionally and privately because of their belief in him, always challenging and pushing him forward. And he looks back with a great sense of pride at the young sports captain who left school aged 15 to grow fruit trees. Each bottle of Man O' War that Matt opens, uncorks decades of memories of life down at the island's bottom end.

HERB FRIEDLI

19

HERB FRIEDLI

–

THE BAREFOOT CONNOISSEUR

Herb Friedli's decision to come to New Zealand was based on a photograph of a tree. He stood on the couch at his friend's place to get a closer look at it; it was a ponga, – a New Zealand native tree fern. His fascination grew. He'd just returned home to Switzerland after travelling through Europe and he was keen to keep going. In 1981, New Zealand wasn't a popular travel destination, so it was perfect for Herb. He jumped on a plane to see that ponga tree for himself.

Herb continued to travel for decades, seduced by cuisine, music, history and his quest for winemaking experience and wisdom, stepping in and out of the old world and the new, before finally settling on Waiheke.

A chemistry graduate, Herb arrived in the Hokianga on the west coast north of Auckland when he was 21. He adored Opononi, where he stayed for a year, earning his keep by collecting seaweed. Further south he was encouraged to visit a fellow Swiss, Hätsch Kalberer, who made wines for Denis Irwin at Matawhero in Gisborne. Hätsch invited Herb to spend the weekend with them on the vineyard. They drank beautiful and very famous wines without any pretension. Herb tasted barrel samples of the Matawhero wines and during that life-changing weekend, he began to understand the possibilities of transforming grapes into great wines. Enthralled and excited, Herb wanted to understand the alchemy of this age-old craft.

Years later, aged 29, Herb enrolled in a postgraduate qualification in Wine Science at Roseworthy College in Adelaide, Australia. It was a three-year course when Roseworthy still had a working vineyard and winery attached to the school. Herb loved every minute out in the vineyard and in the winery. He discovered and started collecting the beautiful wines from Mount Mary in Victoria, established by Dr John Middleton in 1976. Herb completed his studies, announcing to his classmates that he was applying to Mount Mary for a job. Everyone laughed at him because it was well known that no one ever got a job there. It was completely exclusive – no cellar door, no visits, no chance of getting on the mailing list quickly and very little chance of ever finding the wine for sale. Herb had tracked down the wine, so figured he could also secure a job at Mount Mary and, within ten days, he received a phone call inviting him to join them. It says a lot about Herb. He is an eternal optimist and will always see the possibilities in life.

Before he took up his job at Mount Mary, Herb went to work for Schahinger Cooperage, a barrel-making company in Adelaide. He was eager to learn a craft associated with wine that wasn't involved in making it. In sauna-like conditions due to the continual fires toasting the wooden barrel staves, the men work in a line. After each part of the process is complete, the barrel is handed onto the next person to do their part. Each week they rotate positions, to allay boredom, experiencing every operation associated with this desperately physical work. Herb was discovering one of the oldest crafts in the world dating back to Egyptian times. He witnessed the deep sense of pride surrounding him and has never seen such collective pride since. To this day, the process of the ancient craft of barrel making has not changed and Herb is one of only two trained coopers in New Zealand.

At Mount Mary, Herb's aim was to find out how their beautiful wines were made. He worked with Mario, the head winemaker, while John supervised each detail. It was a small operation and they did everything with absolute precision. John's perfectionist approach did not allow for shortcuts. One day, quite abruptly, in the middle of harvest, John halted all work and sent the pickers home. He had found three vinegar flies in the winery and nothing could continue until the source was found. Mario and Herb scoured the winery inch by inch, finally discovering a leaking wine barrel. It was fixed, the vinegar flies disappeared and the following day everything resumed as normal.

Herb's winemaking heart lies in Burgundy and so, during those three years at Mount Mary, he spent the off-seasons working vintages in France. On arrival in Dijon he spent five days grappling with the language before entering his "church" of wine. He simply knocked on doors and they opened - in those days it was still possible. People were very welcoming and, sensing his passion, generous with their time. He tasted unbelievably beautiful wines from both bottle and barrel, wines that transported him, and transcended what he imagined could be possible.

He applied to his favoured vineyards for vintage positions, but he'd left it too late as all the positions were filled. After a week his money ran out – he had nothing left to pay his campground fees in Beaune. In desperation, he turned to his cooperage skills and knocked on the door of Billon Tonnellerie to ask for a job. They accepted him immediately, teaching him even more about coopering and all about speaking French.

Still in search of a vintage position, he took his lunch to sit amidst a trio of Grand Cru vineyards – Richebourg, La Tache and Romanee Conti – for inspiration. He spied a stunning-looking vineyard he'd never heard of called La Grande Rue, producing wine by Francois Lamarche. Shortly after the sacred French lunch hours of 12 until 2, he knocked on the door to ask for a job and the immediate answer was yes. Lamarche is a producer of beautiful wines made in a very traditional Burgundian style, and Herb returned to work there for three consecutive vintages. He collects their wines and has watched the evolution from male to female winemaking, stunning yet different expressions flowing from the wines.

Herb also worked vintage at Chateau Montus in Madiran, South West France. Together with a 70-year-old builder and 200 Roma, he harvested then vinified the Tannat – a variety that was entirely new to him as a winemaker. The experience of living together with the Roma is indelible to Herb and he remembers it as a rare and immensely enjoyable opportunity to experience their culture.

He was offered his job back at Mount Mary, but knew that he had to move on. He had learned to make elegant, nuanced Burgundy and Bordeaux style wines at Mount Mary – wines that were more French than Australian. Seasoned with experience and confident in his skills, he wanted to branch out and experience a little more independence as a winemaker. He'd always wanted to return to New Zealand and Waiheke was the closest quality wine producing area to his beloved Hokianga that he knew of.

In 1994, following harvest, he took on the wine-making duties at Stonyridge. Stephen White passed the reins over to Herb and gave him the opportunity to stamp his personality onto this famed Waiheke Bordeaux blend. He stayed until after the 1997 vintage. Any downtime was spent in the Hokianga, or sharing exceptional wines with friends while honing his impressive culinary skills.

Sometimes you come to a crossroad in your life, when instead of chance landing at your feet, you have to make firm decisions about your future. Herb needed to discover where in the world he wanted to make wine. He flew back to Europe, to uncover the answer. Within a short space of time he felt home-sick for Waiheke – he realised where his future lay.

During the 1990s, many vineyards were planted on Waiheke. One of those was Te Whau, in Rocky Bay where Herb got his first wine-making job upon his return from Europe. He also took on the wine-making roles at Kennedy Point and

Christensen Estate – now Poderi Crisci. It was an extremely busy time in his life. He was one of the most experienced and trained winemakers on the island back then. People would seek him out to help with particular issues. During vintage, he did the analysis for many of our vineyards. Arriving home at 11pm from the winery, he'd find his letterbox filled with samples to analyse. At 6am the next morning after a few hours sleep, he'd ring around with the results. Looking back, he doesn't know how he kept up that regime of 16-hour days for two months solid. Friends would invite him for dinner – not to linger, just to eat and run away into the night.

Nowadays Herb is winemaker at Poderi Crisci. He is training two winemakers on Waiheke in the craft of coopering. He's concerned that the knowledge will die out and feels a duty to pass on this ancient tradition. He remains an incredibly generous mentor to many, myself included. He's more than happy to share his wisdom and his cellar of carefully selected wines that are no longer available or affordable to those of us eager, passionate, interested and who have gained priceless experience from doing so.

While working vintage at Matrot in Meursault, Burgundy, Herb was chatting to father Matrot while his son Thierry was out in the vineyard doing analysis of the fruit, ready to decide when harvest would commence. Father Matrot thought the modern tools were amazing, commenting that prior to this technology, harvest would commence the day after the horse chestnuts fell from the trees lining their courtyard.

On the pre-arranged first day of harvest, Herb walked into the courtyard carpeted in horse-chestnuts. He raced over to father Matrot, amazed at the coincidence of timing. Matrot Senior roared with laughter, acknowledging that the old ways work just as reliably. That was an enduring and valuable lesson for Herb – nature makes trust-worthy and intelligent decisions for you. After all, it was a tree that brought Herb to New Zealand so many years ago. He's very thankful to have learned his craft from the older generations of families whose wisdom he now passes on to the new generations of winemakers lucky enough to cross paths with this gentle barefoot guy. We have benefitted enormously from his sage presence in our wine community.

CAT VOSPER AND BARNETT BOND

—

CAT VOSPER AND BARNETT BOND
MIRO VINEYARD

—

WINE AND THEATRE

The lawyer and the doctor met on stage. He was practising as a GP in Matamata. She was a young lawyer, mad keen on the theatre. Barnett Bond and Cat Vosper were both cast in a Roger Hall play and it is Roger – just quietly, maintains Cat – who is responsible for everything that has happened since then. She has no hesitation in reminding the popular playwright of this fact during his regular visits to Casita Miro, the couple's much awarded Spanish tapas restaurant at Miro Vineyard. And much has happened to this colourful, industrious and delightfully inspired couple since those nights on the stage.

In 1969, Barnett Bond was studying medicine in Dunedin at Otago University. He got a job at the City Hotel and fell head over heels in love, not with a woman – but with wine. His boss, Stewart Elms, who went on to found Felton Road vineyard in Bannockburn, visited Bordeaux annually, buying wine en primeur to send back to Dunedin. So restaurant staff attended weekly wine tastings – a very rare opportunity in New Zealand at that time. Barnett was drinking chateau-bottled Bordeaux before he'd even tasted red wine from the Barossa Valley. The contrast, when he did, was enormous.

Nearly two decades later, Barnett and his friend David Reid started visiting Waiheke to taste wines from the first island producers – Goldwater and Stonyridge. They were very impressed with the quality of the wines and knew it was the perfect location for them to realise their ambition to establish a vineyard to produce Bordeaux-style wine. Business partners Ernie Staheli and Don Locke came on board, providing capital. They purchased a 42-acre block in Onetangi which they named Gulf Crest. They'd all drive up from the farming community of Matamata to spend weekends planting vines. The new fast-ferry services to Waiheke zoomed them over to the island. They had an old Triumph Herald waiting for them in the carpark at Matiatia in the days when you could leave a car there for as long as you liked. The door locks were broken and the key for the ignition was a trusted old teaspoon that worked every time. Now and then they'd find old food wrappers inside, evidence that the Triumph had been borrowed and safely returned in their absence. That was Waiheke back then, drizzled with a hippy-like trust and love in everyone and their possessions.

Weekends were spent working hard, rewarded by total immersion of their sweaty bodies into the waters of Onetangi Beach below the vineyard. Barnett had three girls from his first marriage who often joined them as did the Locke family and the tribes of kids would race around. Their first Gulf Crest vintage, a blend of cabernet and Merlot, was made at Goldwater Vineyard with Kim Goldwater himself at the helm.

In 1990 Cat and Barnett travelled to Bordeaux to soak in the history and culture of their favourite wine style. Barnett was introduced to Cornelis van Leeuwan, vineyard manager of the very famous Chateau Cheval Blanc in St Emilion. Everyone was generous with their time and knowledge, which would soon give Barnett and Cat the confidence to take on sole responsibility for a vineyard of their own. They returned home to Matamata, sold their shares in Gulf Crest in between having two sons, and bought the property next door.

Planting began in 1993 and they moved permanently to the island in May 1994. Just one week after their daughter Miro was born, Cat arrived to her new life. Barnett and the two boys George and Harrison followed soon after. Renovations were still in progress. During that winter, the relentless southerly wind noise of flapping and buffeting echoed through the house engendering a Southern Hemisphere 'mistral' madness. Cat would carry Miro bundled warmly in her cot out into the very steep vineyard while the two young boys spent the mornings at play centre. She'd had three children in three years so life was stretched, and she had to snatch every spare moment to work on the land while Barnett built up his General Practice. She wonders how she survived the isolation and is so grateful for the support of the few friends they had, especially Jeanette and Kim Goldwater.

It was difficult to buy vines in those days – our wine industry was growing quickly and demand outweighed supply. To overcome lengthy queues they grew their own rootstock in Matamata using the extremely fertile run-off from the cow-shed. Varieties were hand-grafted onto the rootstock and 60% of the original vineyard was painstakingly planted this way. A very small harvest in 1996 produced their first vintage. The 1997 and 1998 vintages were made at Stonyridge and in 1999 their first vintage was made on site at Miro Vineyard.

With absolute delight they welcomed into their home Mathieu Griveaux, a French winemaker to assist them in that first vintage on home soil. He was a gauloise-smoking, jazz-playing, practical and unflappable guy, able to make equipment if it couldn't be bought. They lost touch over the years, but one day a patient of Barnett's walked into his surgery to say he'd been at a wine tasting in the Luberon in France, when the hosting winemaker asked about the Waiheke t-shirt he was wearing. That was Mathieu – and Barnett and Cat visited him at Domaine de Tara in the Vaucluse region, bearing a few bottles of the 1999 vintage, which Mathieu proudly shared with his colleagues.

In 2000, a rentable cottage was completed on the vineyard, sailing high above Onetangi beach. The children grew and helped on the vineyard. They'd get lured into jobs that promised fun. Harvesting of the unnecessary small green bunches of fruit meant war games with their loot, aiming and firing at anyone within sight. When birds invaded the nets through unsewn holes, Cat would be waiting at the bus stop at the end of their driveway, enticing the kids to make as much noise at they could with pots, pans and spoons, running up and down the vines, rewarded handsomely with ice-blocks. Winter pruning meant toasting marshmallows on bonfires during crisp winter afternoons. Their teenage years of bonfires were spiked by the enormous booms and echoes exploding through the valley caused by aerosol cans being covertly, thrillingly tossed onto the fire.

Barnett wrote a wine column in the New Zealand Doctor for 22 years until 2016. It was a much loved column written in a humorous and light-hearted tone. Cat used her legal skills to draw up a constitution for the Waiheke Winegrowers Association, which she presided over for the first couple of years. Our growing industry began to understand the need to work together to promote the island's vineyards.

By 2006, wine tourism was gently unfolding on Waiheke. Visitors would walk up the driveway winding around the steep bowl of vines to arrive, unannounced at Cat and Barnett's door, looking for the tasting room. The ever hospitable Cat felt the need to provide food alongside wine tastings, and the little house of Miro was born. Not only is Miro their daughter, Juan Miro is their favourite Spanish painter and the Miro is also a native tree with pinkish purple berries. The name Miro threads the old and new world together, linking wine, food, discovery and culture. In time, Cat banded together with Sarah la Touche, a Kiwi chef with French experience, and they opened a small seasonal tapas bar, converting the vineyard cottage into a kitchen with a small bistro outdoor dining room that could seat just 22 people.

In 2008, a stunning glass pavilion was built to house the growing crowds. The high ceiling floats atop an elaborate system of weights and pulleys allowing the entire mid-section of glass windows to be lowered in summer, opening the restaurant up to the vineyard below and the seascape beyond. Tall cypress trees salute your entry amidst bushes of deeply-coloured roses and the mad sway of hollyhocks. You enter through towering antique doors, made in France and imported from a hotel in Saigon where they spent a century opening and closing on history. Perched above the restaurant sit the gorgeous mosaics of the Bond Bar, urging you up into their colourful nest.

Sales of their wine also grew and they diverted some of their production into Rosé – a much loved summer lunchtime wine. It had the perfect dual effect of directing only the best fruit into their Estate Bordeaux blend, the 'Bond' and their Syrah. Eventually they planted Pinot Gris and Albariño at Miro and they also serve a tiny amount of Viognier produced from a neighbouring vineyard that Barnett manages.

And finally they travelled to Spain, in search of delicious family-produced Sherry. Casita Miro began to host live performances – flamenco, tango, opera, theatre and music. Friends return, their families join them, their children get married on site or work with them – an endless cycle of growth, amassing memories of which Cat is immensely proud. It is not easy operating a restaurant on Waiheke. Power and internet supply can be challenging. You manage your own water supply and sewerage reticulation amid storms and droughts. You grow what you can on site but you rely on supplies to arrive from the city, on time, eventually or not at all. Cat recalls a day when a water main burst during lunch. A table noticing their neighbours' feet underwater, raced out to their V8 Valiant to grab tools. They were drain-layers, and they dug until they found the source of the burst, which Barnett fixed and another day of lunch service dawned. Cat is a fifth generation farmer and her brain is hard-wired to cope: she and Barnett are capable of fixing the distressing problems that arise during service.

The warm and lasting relationships accumulated, nurtured and fuelled with home-grown care is what drives this family. The kids have grown up and left home – but not really. They're very much involved in the vineyard and restaurant, and are bright and loving testament to their courageous parents' industrious, creative and warmly-embracing kingdom.

MIKE WOOD

—

MIKE WOOD
OBSIDIAN

—

WINE AND MIGRATION

The love of a fine woman turned Mike Wood away from arboreal work – in spite of his name – to viticulture. During his European "OE" – overseas experience – in 1994, he met Jeanette while on a short-term arborist job in Baden, Germany. When his job ended, he didn't want to leave her, and so he found work as a cellar hand during vintage. He's never looked back. He soon convinced Jeanette to spend time in New Zealand with him so that he could achieve two goals: obtaining a working visa to enable him to go back to Germany, and obtaining a diploma in viticulture and winemaking in Marlborough. Wine and a woman had redirected his future.

After Mike graduated, he and Jeannette lived my ideal lifestyle: they followed the summer around the globe. Each year, Mike worked six months as assistant winemaker for Simon Gilbert Wine Services in Mudgee, New South Wales, near where his uncle owned a large vineyard. And then the young couple would fly to Germany to a small village near Baden, to work at Wiengut Heitlinger for the next six months. It was all great experience: the large commercial operation in Australia contrasted starkly to the smaller-scale vineyard in Germany. Occasionally, the couple would pop over to Auckland to catch up with family, making the trip over to Waiheke for rest and relaxation.

After three years of moving to and fro, Mike wanted to move fulltime to the southern hemisphere. Jeannette wasn't keen on living in Mudgee, so Mike got a job with David Evans at Passage Rock on Waiheke Island in 2002. He had been craving the hands-on work of a small-scale vineyard and that's exactly what he got: he moved from a commercial harvest of 7000 tonne to one of around 30 to 40 at Passage Rock. David was one of the pioneers of Syrah production on Waiheke and Mike loved working with the variety that had captured his heart in its spiritual home, the Northern Rhône in France. He had to be multi-talented, however: Mike was occasionally called in off the vineyard to help out in the Passage Rock bistro, making coffees and pizzas. Everyone jumped in when needed, as is so often the case with our small businesses on Waiheke.

Mike worked together with David for three vintages before moving to Obsidian as vineyard manager, viticulturist and, from 2006, winemaker. When Mike joined the team, the vineyard was planted mainly in Cabernet Sauvignon and Merlot, with much smaller parcels of Cabernet Franc, Malbec, Petit Verdot and Syrah. All the fruit was sent to Coopers Creek in Huapai, West Auckland for processing, with winemaking overseen by Andrew Hendry, founder of Coopers Creek and shareholder in Obsidian.

Obsidian vineyard has changed significantly since Mike's arrival. They have pulled out virused and inferior plants, replacing them with the superior genetic material available today. Mike's experience of Syrah with David Evans convinced him that it was well-suited to Waiheke, very forgiving and reliable, producing beautiful wines across our vintage variations, and so they planted more. Now eleven varieties are planted on 25 acres, including the Bordeaux five, Syrah, Tempranillo, Montepulciano, Chardonnay, Pinot Gris and Viognier. Obsidian also have two small trial blocks growing varietals such as Nebbiolo, Arneis and Albariño, made into small exploratory batches of wine.

From 2008, the winemaking operation has been Waiheke-based, at Mudbrick's Shepherds Point winery in Onetangi, apart from wine bottling, which takes place at Pleasant Valley Wines in Henderson. Moving the winemaking to Waiheke allows for the flexibility required at harvest time. Still, Mike has very little down time, being both viticulturist and winemaker. He and Jeanette have two children, Natalie and Antony, who are extremely proud of their dad, and Mike makes sure he takes time off with family. He happily holidays in tropical climes without a vineyard in sight.

Many people have been hugely supportive of Mike. When I interviewed him, he'd just been to a Ken Moon tasting of twelve magnificent wines, mostly First Growth Bordeaux, and some Hermitage. Ken's tastings are legendary and to be invited is a rarity. Another supporter has been wine negociant Jean-Christophe Poizat of Maison Vauron, who introduced Mike to the Courbis family in St Joseph, located in the beautiful and impossibly steep valleys of Northern Rhône. This led to a couple of seminal, immeasurably rewarding trips: in 2007 and 2009 Mike worked vintage at Domaine Courbis in Chateaubourg, St Joseph. The Courbis family have been working on their land for over 400 years, and they and the village welcomed Mike into their world, sharing insights built over generations. Mike also visited the entire region, gaining an understanding of the variation of the sub-regions and their wines.

One of the original owners in Obsidian, Lindsay Spilman, worked closely with Mike. Very tough decisions sometimes have to be made on every vineyard, and Lindsay never hesitated in backing Mike's call to walk away from bad fruit. Happily for the winemaker, Lindsay was totally committed to quality. In 2014 on Lindsay's retirement, Alan and Bev Wiltshire took full ownership. They had become shareholders in 2004 when discussions and tastings with his old school friend Lindsay convinced Alan that Obsidian was the perfect opportunity for his family to experience the pride and pleasure of producing wine. Alan, too, never lets the team down: I've made emergency calls to him amidst restaurant summer madness, pleading for immediate wine delivery, to which he always responds. And, in recognition that Obsidian is all about the skills and passion of its viticulturist and winemaker, Mike has become a proud co-owner of Obsidian Vineyard.

Mike has been at Obsidian more than a decade and says he is only just beginning to understand the land, the climate and the effects on the varieties they've planted out. Also a consultant winemaker to many vineyards, he feels very fortunate to be working on this beautiful island in a job that he loves. This softly spoken thoughtful vigneron is a gifted craftsman and even though he says he could live anywhere in the world where great wines can be made, I'd wager he and Jeanette are here to stay.

DAVID AND VERONIKA EVANS-GANDER **PASSAGE ROCK**

—

WINE PIONEERS

On the table in front of us at Passage Rock Winery is a photo of a caravan, a shed, and a yellow Morris Minor car isolated on a wide valley floor. It is 1994 and David and Veronika stand in the frame, surrounded by their 41-acre property in the Te Matuku Valley at the south-eastern end of Waiheke Island. The bare flat land stretches down to the estuary, slipping into the bay, fringed by low-slung peninsulas on either side. At the time, the caravan, the car, the shed – that was it. But Veronika and David have since transformed that valley floor into a family home, vineyard, winery and restaurant - little by little, with absolute dedication.

It's a long way from the Swiss Alps where Veronika grew up, on Camarna Farm in Graubunden. This Canton is probably best-known as the picturesque setting for the children's novel Heidi and is affectionately named "Heidiland" by locals. From a young age, Veronika roamed the steep landscape, rain or shine, listening for the sound of cowbells to find and deliver the herd home for milking. The family were self-sufficient, making their own cheese and other milk products, growing their own food. She has always been drawn to life on the land, incorporating the practise of sustainability.

When David was a teenager, he and his friends would slip across the Hauraki Gulf from Auckland to the family bach on Waiheke and make mischief. In his early 20s, he travelled further afield to London, to work in business, but it was the world of wine that he became interested in. He came back to New Zealand to reimagine his future, visiting his sister Mary who was planting Heron's Flight vineyard in Matakana. He began to explore the possibilities of combining work and wine and was accepted at Roseworthy College in Adelaide to study wine – although he ultimately decided to learn on the job instead, eagerly stepping into the assistant winemaking role at Vasse Felix in Margaret River.

Veronika trained as a handicraft and home economics teacher but was keen to travel and work on farms. She opted for Australia and arrived in Busselton in the South West to work on a Horse Stud. She met David at the annual Margaret River barrel race, where teams raced a wine barrel for a kilometre over land and river from one winery to another. They travelled together around Australia, and before Veronika returned to Switzerland, they visited David's family in Auckland, staying with his other sister Marina on Waiheke Island. Then, David followed Veronika to live in Switzerland; the couple were conscious of the need to understand each other's culture – to decide where they would build a future together.

Graubunden is home to many vineyards, producing mostly Pinot Noir. David found work on a small vineyard and it's where he says that he learned about work ethos. The day began at 6.30am and ended at 6.30pm. Winter temperatures could plummet to minus five degrees. The small team shared viticulture and winemaking duties. They bottled on site, delivering their wine to customers in wooden crates. They were piloting sustainable viticulture practices; David's eyes were opened to new techniques. He tolerated this continental climate, but he and Veronika had their hearts set on establishing their own vineyard, creating a self-sufficient lifestyle, to raise their children in a close knit community. The pull to New Zealand was strong, so they planned a holiday to find themselves a property – and bought in Te Matuku Bay on Waiheke before flying back to Switzerland.

Veronika and David are true pioneers. As they told me their story of living on the land in their solar-powered caravan, I could feel the weight and magnitude of their endeavour. It took two years to build their poured-earth house. David had a job in Auckland initially and the simple chore of ironing his shirts with an invertor, relying on solar power, was a challenge that we laughed about, but it must have been a nightmare at the time. He soon quit the city and the pressed business shirts, and got work locally, training job seekers in horticulture. Veronika worked out at Man O' War in the vineyards, studying viticulture by correspondence. Their first plantings were trained on branches collected locally – a method used commonly in Switzerland, until the posts are finally punched down. The decision was very pragmatic because they simply did not have the finance available to pay for posts.

David then got a job at the local store providing building materials, gardening supplies and home appliances, taking full advantage of staff discounts to buy the many bags of cement to produce the 70 cubic metres of poured earth that became their family home. Weekends were spent making the earth bricks. As the walls rose higher, they used a front-end loader to lift the wheelbarrow up to pour the bricks. The walls took nearly a year to complete, so they celebrated their achievement with friends and family under the open skies. Another year later, the house was ready and they left the caravans behind.

When you are in your twenties, says Veronika, you have endless energy available to bankrole your dreams. Even though it was desperately hard at times, they just plodded on, day after day, week after week, careful not to look into the long-term lest it overwhelm them. Their first vintage was not the happy occasion they anticipated. They just didn't have the labour necessary to overcome the weeds growing happily on their organically-managed vineyard. With no money for netting, the scarecrows could not keep the birds away. So they pressed just over 100 litres of juice and filled half a barrel with wine.

But their lives were peopled with friends and colleagues, sharing advice, meals, equipment, helping each other out. The second vintage in 1998 filled four barrels. The third vintage yielded twenty barrels, crowding the winery room at the back of the house. They needed more space, so in 2000 they viewed and then bought an entire building due for demolition in Auckland. It had been a Garden Centre and it arrived on site complete with original cabling, fittings and fixtures. It was meticulously rebuilt on the huge concrete pad that would become their restaurant, winery and cellar door. They couldn't afford to do it, but they needed further investment to grow the business so they had to make it work. Their major aim was to produce quality wines that were affordable. From the outset, their intention was to win medals and trophies, to secure third-party endorsement of their wines that would persuade the public to visit them in Te Matuku Bay. But after travelling out to the bottom end of the island to taste their wine, the visitors really needed to eat.

A pizza oven was built and word-of-mouth spread about the delicious wood-fired pizzas and award-winning wines on offer. Lovely stories abound of the dashing Robert Hancock, their pizza-oven builder whom they co-opted into making pizzas for a year. Written into the job descriptions of all employees was the requirement to make pizza when requested. In 2001, they not only opened the restaurant, their son Johnnie was born at home. Five years later, their second home-birthed son Marvin completed the family.

The land flourished under their care, planted with vines, trees and gardens. They began to understand their vineyards and what they were capable of producing. Brian and Alison Walker had leased land from them in 1999 to plant Syrah and David began making wine for them, while also consulting and making Syrah for John Bollard on Te Whau Peninsula. They had their own Syrah planted from 1999 and also top-grafted many of their Bordeaux plantings with Syrah. David tapped into his experience of making Syrah at Vasse Felix in a similarly cool climate.

When interviewing for this book, David's name came up constantly when discussing Syrah. The Passage Rock 2003 Syrah won the trophy at the Air New Zealand Wine Competition, followed by many more, placing Waiheke firmly on the map for quality Syrah production and inspiring others to plant it on Waiheke.

Investment in the winery grew with the installation of white wine-making facilities and a bottling plant, which other vineyards also use. David also took on the management of additional vineyards around the island to grow their volume and facilitate the affordable pricing of Passage Rock wines. At Passage Rock they grow Syrah, Cabernet Sauvignon, Cabernet Franc, Malbec, Chardonnay, Viognier, Pinot Gris and Sauvignon Blanc. They have seven hectares planted on their home vineyards and they manage eight additional hectares in Oneroa. It's the perfect equation for growing volume without having to purchase land.

A highlight of the last two decades for David and Veronika has been the staff who have come and gone. It's not a big company, so there isn't a long ladder for employees to climb. David remembers when he first got into the industry, relishing the collective wisdom and energy in the winery and on the vineyard. He's spent many years replicating that feeling for the young wine-making assistants coming through, gaining the experience they need to step up to the next level elsewhere.

We talk about the future – of what it may hold for them. Veronika is redirecting energy into her passion for arts and crafts. She plans to visually create awareness in visitors to the connection between the land and the wines they are drinking. She understands clearly that the aim in life is to be true to yourself, to reinterpret your dreams as your life develops and changes. "You have to be authentic – that's the challenge". This couple have worked so creatively and bravely to build Passage Rock Winery. And when you visit, imagine that bare, wide valley floor all those years ago, stretching out to the shoreline and marvel at their dedication over the passage of time.

TONY FORSYTH

—

TONY FORSYTH
TE WHAU

—

1 + 1 = WINE

Tony Forsyth's passion for what he calls "a very civilised and very civilising convivial beverage" is palpable. If you are fortunate enough to spend time perched at his bar, he will share the most remarkable wine memories and experiences. They should have satisfied him for a lifetime. Instead, with his patient, smart and supremely committed wife Moira, he decided to plant a vineyard, make wine and open a restaurant. There have been moments that nearly killed him with stress, but the establishment of Te Whau has been immensely rewarding, enabling him to share his wines with visitors from around the world. The only other subject upon which he may engage you with the same intensity is Point Theory mathematics, where 1+1=1. If you'd prefer to slink away, glass in hand and quietly watch the sun go down, he won't be the least bit offended.

Tony lived in London in the late 70s and early 80s. His boss was a wine enthusiast who shared his passion and knowledge. Tony would escape to France and cycle slowly through all the wine-producing regions to understand the terroir, the varied landscapes and their wines. He bought his first bottle of Calon Segur, a classified Bordeaux Third Growth, for only 14 pounds. He was smitten.

Back home to live and work in Auckland, he continued collecting wine – Bordeaux, Burgundy, Rhone and our own top Waiheke wines at that time: Stonyridge Larose, Te Motu and Goldwater. He had a well-paid job and didn't crave fast cars or fancy houses – just great wine and excellent food. He'd trained as a psychologist, working as a consultant recruiting and placing professionals in top jobs. Before he'd turned 50, he'd climbed the corporate ladder to become Managing Director of Sheffield Consultancy Group. He'd met the local wine producers and became intrigued by the possibilities of quality wine production on Waiheke. He reasoned that if they could do it, maybe he could too.

Moira, knowing his intention wouldn't falter, suggested he go and find the right piece of land and get on with it. The only proviso was that he had to plant some Chardonnay and make her favourite wine. In 1993, on a block at the end of a new subdivision in Rocky Bay they struck gold. Steep north-facing slopes rose high above the sea, with incredible vistas back to Auckland. It was a financial stretch: Tony remembers standing on the top of the hills, friends shaking their heads and telling him that it would never work. The slopes were too steep. But he'd seen the initial plantings at Man O' War, not to mention the vines clinging to hillsides throughout Europe. He'd jumped ship from financial security to embrace an island life - he just had to find a spectacularly talented tractor to negotiate those slopes.

Tony gives full credit to our early pioneers – Goldwater, Stonyridge, Peninsula and Te Motu. "They didn't know what they didn't know," he says, and they could only find out by doing. He feels very lucky to have come along in their wake, profiting enormously from their experiences. His degree in psychology, with minors in chemistry, microbiology and botany gave him a science base upon which to build his business and a psych-trained inner voice to boost morale when needed.

The major headache of planting a vineyard and opening a restaurant in 1996, was compliance with the then-new Resource Management Act. It was a bewildering process, yet together with Moira – who, Tony concedes, actually does all the hard work – they persevered, transforming 30 acres of land into vineyards and creating 12 full-time jobs.

The vineyard is planted with 50% Cabernet Sauvignon, 30% Merlot, 15% Cab Franc, 3% Malbec and 2% Petit Verdot, constituents for their blend called The Point. They have a Chardonnay block for Moira – although they produce so little of this sought-after wine, it's nigh impossible for her to lay claim to her fair share. (Hopefully she didn't notice Tony and I sipping our way through a bottle during the interview!) There is a tiny, super-steep block of Syrah. The initial winemaker was Herb Friedli and the wines have been critically acclaimed from the outset. Many famous palates have visited, including best-selling wine writer Hugh Johnson, who named Te Whau in his book of 200 favourite wineries in the world to visit.

Tony has had some mesmerising wine experiences. One was a visit to Ramonet in Burgundy with the President of the Sommeliers Association of France and the President of the Burgundy Producers Association. They knocked at the enormous door and a tiny window opened to reveal a face peering out at them. They descended to a traditional cellar, cobwebs intact, and tasted wines from every barrel, finishing with a dusty bottle plucked from a rack. It was a 30 year-old bottle of Le Ramonet, a world-famous Chardonnay. "It was still crisp, with layer upon layer of flavours opening up", remembers Tony.

He had lunch with Monsieur Comte Alexandre de Lur Saluces at Chateau Y'Quem in Graves - home to the famous, absolutely luxurious dessert wine of the same name. Alexandre showed him where 17th century oak panelling had been removed from the original Chateau to reveal astonishing 15th century frescoes, fresh as the day they were painted and unseen for three centuries.

Tony was born in New Zealand but lived in England as a boy and the family visited many cultural sites throughout Europe due to his father's deep admiration for art and architecture. Te Whau's site is sublime and demanded an impressive building. A visit to Le Corbusier's Notre Dame du Haut in Ronchamp in 1959 haunted Tony, inspiring the eventual design of Te Whau's three-storied tower. The cellar at Te Whau is dug into the hillside – a secure and temperate fortress for wine and a sanctuary for elevage. Te Whau built one of the first gravity-fed wineries on the island. Incoming fruit from harvest is trucked to a sorting table on the first floor then fed through a manhole into the top of the fermentation tanks on the ground floor below.

The restaurant floats above, with floor-to-ceiling windows offering panoramic views across the Hauraki Gulf and islands. The tasting room walls are lined with wine books and a crowd of drained wine bottles. Rows of famous and superbly aged labels face the room, causing wine buffs to dribble with delight and desperate envy. Tony is incredibly generous, and many of us have been the lucky recipients of phenomenal wines from around the world. He'd host intimate dinner parties mid-winter, with his chef creating delicious adventures on the plate to match the phenomenal wines selected from his cellar. These incomparable evenings taught me about world-class wine, but also helped me to understand why it is that people remain so desperately under its spell.

Tony has achieved a great deal, but declares outright that opening up the restaurant was the most stressful thing he has ever done. He had no experience in hosting and managing a restaurant floor whatsoever. I can verify this terrifying fact, as Tony employed me as his sommelier. It was a complete privilege, albeit a frightening opportunity, to understand and recommend his vast wine cellar. But it was also deeply distressing to discover that Tony had booked all of the tables on opening day to arrive at the same time. How on earth, I asked him, would the kitchen cope? He raced to the phone, mercifully arranging staggered arrivals and, for the first time, money trickled in instead of pouring out of the restaurant.

The restaurant has hosted a vast array of guests, some very well-known: Patrick Leon from Chateau Mouton Rothschild, Sean Connery, John Cleese, Gordon Ramsay, Jamie Oliver and Martina Navratilova to name but a few. It certainly contributed to placing Waiheke firmly on the international destination map. Yet Tony the psychologist hadn't realised how interesting and challenging the public can be. The former visitor became the oft visited, and he's found it tough to overcome the emotional impact left by customer behaviour and comment. I recall an irate customer throwing potatoes across a crowded restaurant because the dish he desired had sold out.

Up to 12,000 people dine in the restaurant annually and another 10,000 visit the cellar door, or just come to have a look.

It may have been over almost before it gained momentum. Grape farming is hostage to the weather and in 2001 nature was cruel to our vineyards. The team at Te Whau made a deeply depressing decision as none of the fruit survived the onslaught of the weather: the entire year's work would be picked and dumped. Surveying the desperate situation after a sleepless night, Tony considered his future, and was uncharacteristically despairing enough that he even contemplated ripping all of the vines out. It was only his third harvest and remains the worst growing season Te Whau have experienced. But this psychologist knocked himself back into shape and the vines stayed put.

The Vineyard Manager, Steve Hood, has been on the team since day one. The winemaking was taken over by Curtis Staples, who joined Te Whau as a young boy out of school, running errands, picking up deliveries, assisting on the vineyard and in the winery. He has a natural palate and went on to study oenology and viticulture while continuing to work at Te Whau. Tony has prized watching Curtis develop into a skilled winemaker, acknowledging that he had a fantastic mentor in Herb Friedli. Curtis has recently joined the talented team of winemakers at Villa Maria, sharing his finely honed skills with the wider wine community. Tony tries to keep out of the way, always present to observe, discuss, learn, participate and work his butt off in the restaurant with Moira.

In another life, Tony would make a great restaurant reviewer because he's learnt the hard way and knows what it takes to produce a memorable food and wine experience. The Danish philosopher Søren Kierkegaard wrote that "to dare is to lose one's footing momentarily. Not to dare is to lose oneself". Along with all of the characters in this book, Tony and Moira Forsyth dared, and we are all the benefactors of their vision.

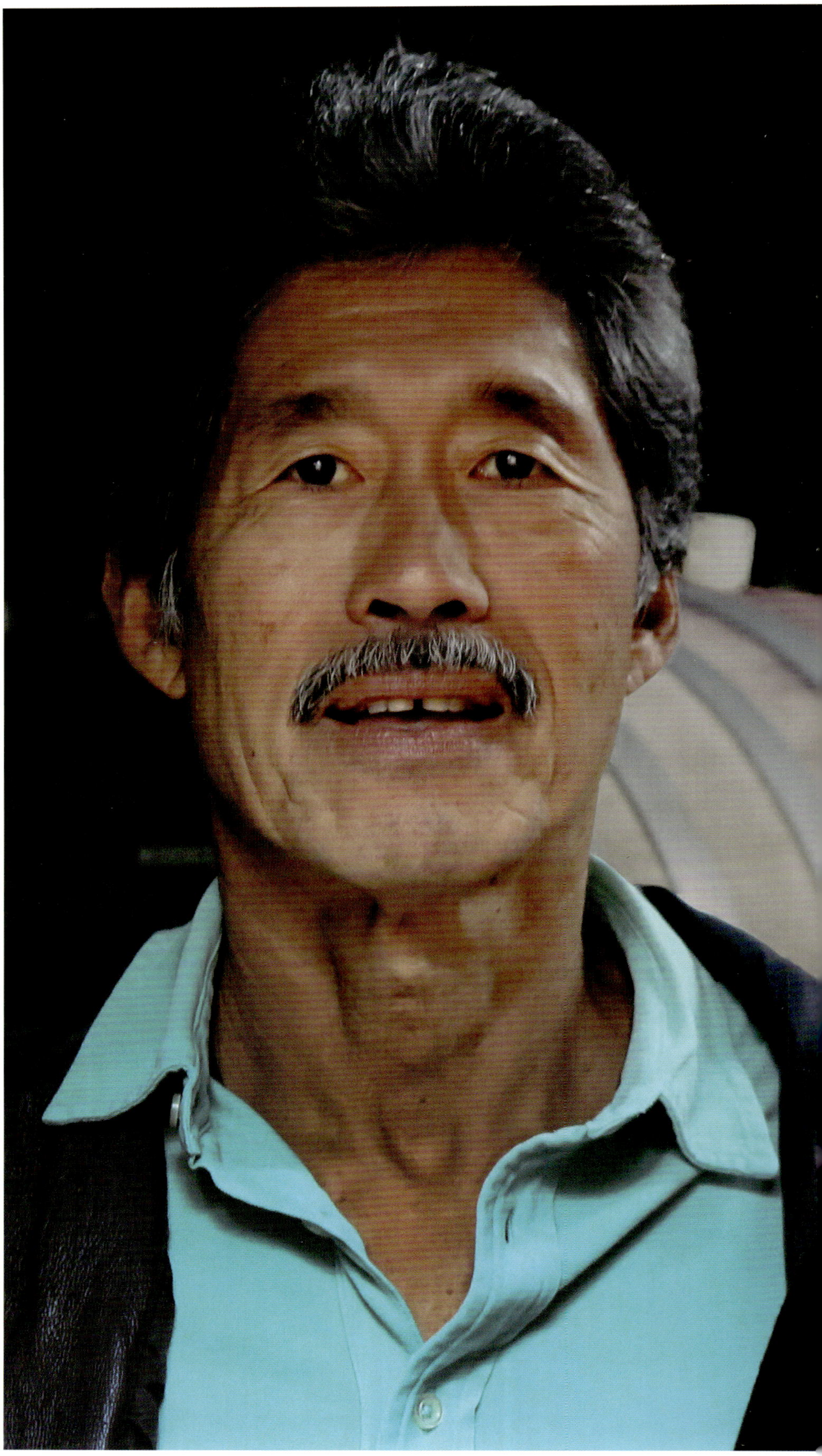

NEAL KUNIMURA, SUSAN MCCARTHY AND ELIZABETH ZHONG

KENNEDY POINT VINEYARD

—

A CULTURAL FERMENT

Our wine community on Waiheke is incredibly diverse. This small island has drawn to its shores people from all over the world with a single challenging dream – to make great wines. They bring with them their rich experiences, both in wine and in business, to grow and enhance our industry. It's not always easy to live out your dream, no matter how ambitious or crazy it seems, but your problems and your successes can be shared when you live amongst a sympathetic community like ours.

We have all learnt from each other. I have been incredibly fortunate to learn about wine by drinking the great wines of the world – straight from the cellars of generous friends like Neal Kunimura and Susan McCarthy. They introduced me to wine experiences one dreams of – imagine being presented with a 1959 Chateau Margaux, followed by a 1961 Mouton Rothschild, to accompany the tender beef rump one Friday night at their lovely home, deep in the heart of winter on Waiheke. And it didn't end there! But that's another delicious story. Needless to say, Susan and Neal adore great wine, great food and sharing their collection and culinary expertise with friends.

It all began back in Hawaii, where they lived. In 1974, Neal started his own printing business, which grew into a successful enterprise. Susan worked in business and sales with a multitude of organisations. A distributor for the famous Mondavi Wines was a friend of theirs, and he shared his knowledge and passion for food and wine with them. One thing led to another until Neal's wine collection necessitated a purpose-built cellar. So they built a new house to create that cellar. But Neal still wasn't satisfied: such was his admiration and dedication, that he had decided to learn how to grow and make great wines of his own. But where?

In 1994, on the advice of Kiwi friends, they travelled throughout the South Island, attending the Marlborough Food and Wine Festival, and then investigated Hawkes Bay. A few days before departure they still hadn't found anywhere to build their dream vineyard, so those same friends suggested they take a look at Waiheke. They briefed the real estate agent who spent the day showing them properties suitable for grape growing. Nothing appealed. Finally he took them to a property he didn't think they'd find appropriate, but which was the last ten-acre listing he had. He drove them up the curving driveway, under a grove of spectacular Pohutukawa trees that paraded down to the shoreline below. Neal wandered over the property, up to the hill-tops falling into the sea, striding back down with his mind made up. The utterly surprised agent was made an offer on the spot, and Susan and Neal flew back to Hawaii to begin the 18-month-long process of immigration.

In June 1996, residency was granted and their wine tourism business sprung into life. Their Kennedy Point property is magnificent and Neal had sensed straight away that it was special. The previous owner spoke of his sheep suffering from facial eczema on neighbouring properties, but never when stationed on Neal and Susan's land. 'Keai' means 'special place' in Neal's culture – a spiritual place, and that's exactly what Neal felt he'd found when he first walked around the property.

There were a few other very influential and important mentors for them in their new venture. Neal did his own research on the planting of grapes but he needed to learn about making wine. One day while Neal was planting, Waiheke winemaker Herb Friedli called by while out walking his dog. He was keen to know what they were doing and, after talking about their project, Neal asked Herb if he could work with them. The seeds for a great friendship were also planted: Herb has a knack of turning up on really beautiful and perfectly-sited vineyards, destined for making great wines, just when someone is in need of a winemaker with his insightful experience and expertise. Neal also credits Hätsch Kalberer from Fromm Vineyard and Hans from Hans Herzog – both in Marlborough. The sum knowledge of these three grape growers and winemakers has been very beneficial for Neal and Susan. Through them, Neal and Susan developed a passion for the wines of Burgundy, travelling to the source and eventually importing the great wines of Domaine Ramonet as a side business, mostly, says Susan, to ensure Neal's cellar was home to these beguiling old world Chardonnays and Pinot Noirs to share with fellow wine enthusiasts.

Initially Neal and Susan planted Cabernet Sauvignon, Merlot, Malbec and Cabernet Franc, the ingredients for their beloved wines of Bordeaux, which Waiheke was becoming well-known for. They thought it would be interesting to plant Syrah and found that it absolutely thrived. They enjoyed the wines of Cote Rotie, where Viognier can be added to Syrah, but had to wait a year before Viognier became available in New Zealand for planting. Every year Neal kept planting more Syrah to increase volumes. Much later they planted a small block of Chardonnay. Neal and Susan love to drink Rosé and produced their first in 2000, when it was just not a sexy wine, attracting very low esteem in this country. They persevered and now, of course, cannot produce enough to supply the soaring demand. After much hard work, the vineyard is certified organic and they use biodynamic preparations. Time has revealed that the vines have a fantastic ability to resist disease and withstand challenging vintages.

After a few years making wine, some extra special vintages gave them barrels of Bordeaux varieties, so beautiful on their own, that they decided they didn't want to blend, but to bottle them individually. They found that visitors to the tasting room enjoyed tasting each variety and being able to learn about the differing characteristics. Neal and Susan decided to continue bottling individual varieties as well as blends, providing more options and experiences for their growing market.

They also use grapes from four additional vineyards on Waiheke. Three of them are managed by their viticulturist Heike Sonnenschein and the fourth is managed by Wendy Saward. Their Waiheke single varietal labels include Syrah, Cabernet Sauvignon, Merlot and Chardonnay. They also produce a Bordeaux blend and a Rosé. In addition, they have a Marlborough Sauvignon Blanc, grown by their friends who introduced them to New Zealand in the first place. They are excellent grape-growers and Neal credits the winemaking skills of Bill 'Digger' Hennessey for the Sauvignon Blanc. With that great addition to their portfolio, an American distributor took on their wines, and they also export to Japan, China and the UK. There are only a few Waiheke vineyards able to grow Sauvignon Blanc, but it's very important to have one in your portfolio because it is synonymous with New Zealand wine, enjoyed and revered around the world.

They spent years planning the business, deciding upon a multi-faceted approach to embrace the growing tourism and hospitality industry on the island. Previous experience in business ensured a very successful strategy. They built a guest-house to accommodate visitors, they planted olive trees for the production of single estate oil, and they created beehives for honey. Susan says there's a lot to be said for vineyard owners who "put their feet in the dirt and go to work": "none of us on Waiheke have just sat back sipping wine". In the tasting room, the main focus is wine, but it's lovely to walk away not only with wine, but also with honey and olive oil to share with friends and family. This beautiful vineyard was made possible by 20 years of hard work, of business acumen, of the gentle spirit of Neal and the clarity of vision and confidence-inspiring presence of Susan.

When I started this project, Neal and Susan were the owners of Kennedy Point, but they have now sold their business, turning towards a more relaxed lifestyle. Together they have provided our small island with exquisite experiences and we are all, in our own way, the memories they will cherish, just as we shall remember and celebrate their legacy.

Viticulturist Heike Sonnenschein, who joined the team in 2000, continues to provide expertise and continuity to the new owner Elizabeth Zhong. Elizabeth is a Chinese-born Kiwi who grew up in Shenyang, in northern China. In 1997, after many years working in Beijing, she immigrated to New Zealand with her family. Elizabeth was first introduced to wine through books. As a dreamy young romantic, she feasted on western literature and arts. Her move to New Zealand inspired a deep passion for wine and Elizabeth knew she had to become involved. Coupled with a love of natural sciences, it was the mystery of wine that captured her heart. After many visits to Waiheke, her growing love of wine and the island collided with the option of buying Kennedy Point Vineyard. The certified organic wines and the viticultural expertise of Heike sealed the deal.

Elizabeth also recently purchased Carrick vineyard in Bannockburn, Central Otago. Steve Green, the previous owner, planted the vineyard in 1993. He has been an inspirational mentor to Elizabeth, sharing his extensive knowledge about the New Zealand wine industry. He's been Chairman on the Board of NZ Winegrowers, with hands-on experience of transforming Carrick into a certified organic wine producer. With Carrick and Kennedy Point, Elizabeth has an enviable portfolio of quality organic wines, showcasing regional diversity.

Elizabeth has become a grandmother and divides her time between Auckland, Waiheke, Wellington and Central Otago. Film-making is her other passion, which she hopes to realise someday. But she is committed to retaining the fine reputations already established by her vineyards. A new, exciting chapter for Kennedy Point has begun.

STEVE POLETTI

—

STEVE POLETTI
AWAROA

—

WINE AND COSMOLOGY

In the mid 90s, a group of eight friends bought 120 acres above Awaawaroa Bay midway down the island, including Steve Poletti and his brother Nick. Both had a love of wine and Waiheke. As boys they'd sail with their family to the island, anchoring in Putiki Bay to visit Kim and Jeanette Goldwater on their vineyard. The families met when they all lived in the United Kingdom and Steve feels like he's known Kim and Jeanette forever. Recurring visits to Waiheke eventuated in permanent residency at Awaroa for Steve. This physicist-turned-economist walked quietly into the world of grape-growing and winemaking, dedicating every spare moment to his quest. Good-humoured, a bit shy and very humble, Steve has spent his entire life in academia. So perhaps winemaking – with its mix of gruelling physicality, science and sensuality – is the perfect complement, engaging both the body and the mind.

Steve's father, Alan Poletti, was a professor of nuclear physics; Steve and his two brothers were born in Oxford when Alan was completing his PhD there, and then the family moved to California while Alan was furthering his academic career. Steve's sister was born there and the family returned to New Zealand when Steve was six. In his teenage years, they travelled again, spending a year split between Oxford and New York City. On their return, Steve was interviewed for his final years at Auckland Grammar. The Principal asked him what he wanted to become. "An astrophysicist", he replied.

Things have turned out slightly differently, although he is indeed qualified as an astrophysicist, studying in Auckland, Canberra and Newcastle-upon-Tyne before conducting post-doctoral research in Adelaide. In simple terms, he was thinking about the origin of the early universe, doing equations and questioning our existence. Philosophical questions that wake you up in the middle of the night. How did we get here? Are there lots and lots of different universes with different constants and slightly different laws of physics? Did we just find ourselves in one universe that allows life to exist? As a physicist, he tells me, it's your quest to find answers to these big philosophical questions. But I was very relieved to shift our focus onto wine.

While Steve was in Adelaide, he also started thinking about wine. He'd take his bicycle on the train to the many wine regions surrounding Adelaide, tasting, learning, growing his love of the brilliant Syrahs and Chardonnays that were his preference. He shared a house, so with very limited storage space, the boxes of wine he bought became furniture, draped with blankets.

In 1995 he moved back to New Zealand permanently, settling into a room in the old farmhouse he and his friends had bought at Awaawaroa Bay on Waiheke. And he made a major career change, knowing that if he'd stuck with mathematical physics, he'd have no choice but to do research work internationally. He'd always been interested in economics from a political perspective and enjoyed discussions with his fellow shareholders in the farm. At the same time, he wanted to contribute to society the way he knows best – that is, academically. He began studying economics, eventually gaining his second PhD, going on to do research in energy economics before being appointed Lecturer at University of Auckland. It's an exciting time in the world to be teaching energy economics, he says. He hopes the future will see the world leave behind its reliance on fossil fuels to use only zero carbon energy. He delights in the energy and enthusiasm of his students. Meanwhile, the Awaawaroa Bay shareholders charter was very flexible, requiring only that the land be organically farmed. Steve and Nick tasted the products of the flourishing local wine industry and one day Nick suggested that they plant a vineyard. Steve loved wine, sure, but he had no idea of the consequences he'd face when he agreed that it was a great idea. Nick was living in Sydney, so realistically, that left Steve to manage their idea.

A group of friends gathered at the farm one weekend to plant out the vines that had been in cool storage. Steve had never seen a dormant vine and wasn't really sure how deep they should be planted. He called on his friend David Evans from Passage Rock, who advised him, and David and Mike Wood have been invaluable teachers ever since. Armed with spades, they spent the weekend hand-digging in the vines. The planting of Bordeaux varieties and a bit of Chardonnay continued for many months. The farm had plenty of pine trees destined to become vineyard posts. Steve and Nick spent long sweaty hours felling trees and one nearly killed Steve as it crashed to the ground beside him. The hand-made posts were sent to town for treatment. When the bills came in, they realised they could have just bought the posts readymade. They were learning on the job.

Years were dedicated to working weekends on the vineyard. Steve is ever thankful to his partner Rebecca Potts, a much-loved local doctor, and their fellow shareholder Alex who commuted to Waiheke most weekends to help manage the vineyards with Rebecca and Steve. Steve maintains they loved it, enjoying the physical work outdoors in contrast to their professional environments. In 1999, with great excitement, shareholders, family and friends picked a very small harvest, which produced just four bottles of wine. The bottles were to be stored for a few years while Steve and Rebecca went to Australia, but outrageously on their return, the bottles couldn't be found. The culprits certainly were, and Steve will go down in history as having never tasted an aged drop of his very first wine.

The second vintage in 2000 made half a barrel of wine. They had no destemmer, so when the fruit came in, friends had to hand-pluck the berries from the stems. In 2000 they planted Syrah, which now comprises half of their 1.75 acre total planting. Steve had tasted a Syrah produced for Brian and Alison Walker on land they leased at Passage Rock and was enamoured with the wine. Awaroa were early adopters of this now widely planted variety.

When some of the shareholders pulled out of the vineyard, Steve, Nick and another partner Barry established a separate company. In 2010 they appointed vineyard manager and viticulturist Matt Sloane, who has done a brilliant job for them, easing the pressure on everyone. The vineyard is planted out with 50% Syrah, 35% Cabernet Sauvignon, and some Merlot, Malbec and Cabernet Franc. Awaroa Estate produces a Syrah and a Bordeaux blend. In a great year they release the Awaroa Reserve wines – a Syrah named Melba Peach and a Bordeaux blend named Requiem. If the vintage tonnage allows, they produce The Dan, a Syrah/cabernet blend. They also release Waiheke Road wines, made from fruit sourced elsewhere, generally Chardonnay and Sauvignon Blanc.

Awaroa celebrates an en primeur day in November each year. Close to one hundred people visit the vineyard to taste the wines directly from barrel and place their order. The following year in November they return to pick up the previous order and taste the new vintage. It's a wonderful way for Steve to get feedback, with everyone comparing and contrasting previous vintages. They fire up the large outdoor pizza oven, feasting on pizzas, cheeses and breads. Unsurprisingly, Steve's customers are very loyal, returning year after year. The lively social occasion yields both financial and psychological endorsement and is a unique opportunity for buyers of Waiheke wine.

Awaroa Estate is carved into gentle slopes surrounded by native bush. The farm is neighbour to a Forest and Bird Reserve, a Regional Park and covenanted land. In spite of electric fences around the vineyards, wild boars developed a taste for grapes and they started slipping in and stripping berries off bunches. It was never enough for serious concern. But during the 2015 harvest, that changed. Matt and Steve's practice is to pass through the vineyard little by little to assess and select individual bunches, to ensure maximum ripeness. They'd picked about half of the crop and decided that the remaining half could stay out for another week. A few days later, Steve was out checking on the fruit – but he couldn't see any. Scattered berries lay here and there, but half a tonne of Syrah grapes – half of their total Syrah harvest – had disappeared. Entire bunches had been bitten off and swallowed whole by wild boar. Steve has always had concerns about the tiny quantities of fruit coming off their organic vineyard, but this was unbelievable. We visited him in the winery shortly afterward and his sage acceptance of the disaster that had befallen them was admirable. One year's worth of careful management had disappeared overnight.

Shortly afterward, a neighbour shot a boar weighing in at 160kg. The council are now regulating the amounts of boar roaming in Awaroa's back yard. Steve is probably going to have to fence his vineyards. But imagine local restaurants specializing in wild boar reared on organic grapes. Matched, of course, with a thimble-full of the 2016 Awaroa Reserve Syrah.

In the summer of 2017, Awaroa opened a tasting room on the farm, just off the main road. The economics of owning a vineyard can be very marginal and it's a great idea to sell your wine direct to the public and plough the entire margin back into the business. The vineyard yields, at a mere 2-3 tonnes per hectare, are possibly the smallest on the island. The establishment of this organic vineyard has been a labour of love. At the end of the day, Steve set out to make wine he loves to drink. He's happy to sit and think about a wine he's tasted for hours – just as he used to contemplate the origins of the universe.

LOUKAS AND CAROLINE PETROU

—

LOUKAS AND CAROLINE PETROU
CABLE BAY

—

A WINE ODYSSEY

Cyprus, jewel of the Eastern Mediterranean Sea, has a winemaking history dating back to 4000 BC. Loukas Petrou grew up in the Cypriot village of Piyi where his father was a builder and his family grew grapes and made wine for themselves, as generations before them had done. They had a small hand press, transferring the juice to amphorae for fermentation. Inheritance is not only about the things or money your parents may leave you. It can also be about absorbing their abilities and dreams and expressing them in your own life. On an island far from Cyprus, this is what Loukas has done, by building a life for himself and his family in our small wine community.

Loukas, his sister and two brothers were all born within four years. The family were self sufficient, as most Cypriot families were in those days. They had fruit and olive trees for oil production. Vegetable gardens supplied their kitchen and chickens ran about. But in 1974 when Loukas Petrou was just sixteen, everything changed. Sadly, his mother died. Then, just one week after burying their mother, Loukas and his family had to flee their home, running for their lives: Turkey had invaded Cyprus. They found shelter with many refugees under orange and lemon trees. They never returned home to the village where they had such deep roots and where their family had lived for generations. They had no choice but to move to the free part of Cyprus and begin all over again.

Two generations beforehand, also on Cyprus, Caroline's grandfather had also said goodbye to everything he knew: his family, the stone fruit orchards, the vineyards – the difference being, of course, that this goodbye was voluntary. In 1930, from the village of Palehori, famous for its wine, he journeyed towards Australia in search of a better life. But he never made it to that Great Southern Land: en route, his ship docked in Wellington, and he walked into a warm and welcoming Greek community and decided to stay, settling in Kilburnie.

The mountain farmer became a coastal fisherman, starting up his own business, gradually acquiring fishing boats. He was the first to export our crayfish to Melbourne. He arrived in New Zealand with nothing and through sheer hard work, brave decisions and persistence, he created a secure and rewarding life for his family. Caroline remembers him sitting her on his knee and telling her the old stories – but she was too young then to appreciate it, eager to continue playing with her cousins.

About the same time, on the other side of the world, Loukas and his siblings found work to support the family into their new life as refugees. He worked in cafes and restaurants – a job he loved. He carried on working in hospitality while in Athens, studying engineering. He took a break from his studies, travelling through Europe and working his way up to become a Silver Service Waiter. He worked in the famous Chateau Briane in Paris. Then, back home one year for the Christmas holidays, he met a Kiwi girl travelling through. She continued on to Athens with him and soon they were on a plane back to New Zealand. Loukas had always wanted to finish his degree in an English-speaking country and the idea of this mysterious tropical paradise appealed.

So when he arrived in New Zealand, naturally he turned to hospitality to pay his way through the completion of a Bachelor of Engineering. His first job was at Top of the Town, a hotel near the University where he was granted a work permit on the strength of his 'silver service' experience in Europe. His English improved rapidly. Even after entering the workforce as an engineer, he continued to work by night as a waiter in a world he knew and loved. He particularly recalls his job at the infamous White Heron in Parnell, serving Rod Stewart and Prince Phillip amongst many more from the international jet set.

Loukas married the woman who brought him to New Zealand and together they returned to Cyprus where he set up an office, taking on engineering jobs. But life didn't work out for them there, so they returned to New Zealand to try and save their marriage. When his daughter Nikki was one and a half years old they divorced, and Loukas stayed in New Zealand to be close to his child. When Nikki was eight years old she went to live with Loukas until she left for university to study law.

Caroline has been a part of their lives for 22 years. Her father, a businessman, moved to Auckland when she was young and the City of Sails became her home. She worked in retail – an extremely busy and demanding environment, which was to provide her with the many skills necessary in her future destination vineyard restaurant, Cable Bay. When the couple met, Loukas noted Caroline's beautiful face and asked if she was Greek. Their following conversation revealed the lovely coincidence that they were both deeply connected to the island of Cyprus.

Now, they visit Cyprus most years; their community there is small and closely bound. Loukas' father, now in his late 80's, still makes olive oil for the family, all of whom live on Cyprus except Loukas. During one trip, Loukas and Caroline visited her grandfather's village unannounced to enquire about her family name. A little old lady dressed in black pulled them inside, made them coffee, fed them and talked about her family. On the fridge door were photos of girls in Wellington Caroline knew – it turned out that little lady was their grandmother. Caroline was able to visit her old family home, to stand in the place where her ancestors had lived for generations – a powerful and emotional experience.

Loukas completed a two-year Masters of Business Administration degree while working with a large construction firm, with the aim of setting up his own business. However, he talked to the owner of a construction firm wanting to sell who advised him that it was a better idea to buy an existing business. He suggested that Loukas work with him for a year to decide if it was the right fit. In 1996, Loukas became the third owner of Canam Construction, established in 1955. His company has grown from 30 to 160 employees. Canam's offices in Auckland are green star-rated: Loukas explains that construction involves "huge capital expenditure and so you must respect and deliver projects, satisfying requirements on every level, including awareness of the impact your work has on the environment". Not coincidentally, Canam Construction built Cable Bay. Loukas knew the owners and worked with them to create their vision. Many years later, he stepped in, eventually taking full ownership of the business.

Cable Bay grow Chardonnay, Viognier, Pinot Gris, Syrah, Merlot and Malbec. They've also planted a small amount of marsanne to blend together with Viognier to create their own Northern Rhône style white wine. They do not buy contract-grown fruit – their aim is to produce wines from the vine to the bottle, with every step of the process under their own quality control. Chloe Somerset is the winemaker, with vintage experience gained here and in Australia, Burgundy and Santorini. Viticultural care of their Waiheke vineyards is managed by Sam Theed, with a deep connection to the vineyards going back prior to the new ownership. Cable Bay own a couple of vineyards in the Awatere Valley in Marlborough, planted in Sauvignon Blanc, Chardonnay, Pinot Gris and pinot noir. The vineyards are managed by Bart Arnst, one of this country's leading organic grape growers. The wines are processed and vinified by Chloe at Mahi winery in Marlborough.

The Cable Bay Restaurant dining room has always been a very classy and upmarket, special occasion experience. Loukas and Caroline wanted to change and expand upon that perception. They want Cable Bay to welcome everyone – whether dressed in shorts and jandals, or dripping in diamonds. They built the Verandah Bar to provide a casual dining experience. The large shaded outdoor dining area offers Mediterranean flavours. A wood-fired pizza oven is also home to slow-cooked meat. There are mezze platters to graze on and beanbags are scattered across the never-ending stretch of green lawn with wide, spectacular coastal views back across the Hauraki Gulf islands to the city. The sunsets are incredible.

The cellar door is located nearby the winery so that guests can smell the ferments and view the winemaking during vintage. There's a small private dining room in the Wine Library presided over by a chandelier, and a boardroom caters to corporate away-days.

Caroline and Loukas enjoy spending their weekends working alongside everyone. Caroline insists she'd really miss it if they weren't part of this large, busy and bustling operation. You will find Loukas wherever he's needed – his love of hospitality and the necessary teamwork it entails has him washing dishes, unblocking toilets, mopping floors, chatting to customers. Often you will find him behind the bar, polishing glasses, listening to the everyday experiences surrounding him, looking out across the blue and green hues of the sea so reminiscent of the childhood home he cannot go back to. This is his Mediterranean island of Cyprus, Waiheke style.

MIKE, ANNE AND SEAN SPRATT

—

MIKE, ANNE AND SEAN SPRATT
DESTINY BAY

—

WINE AND OPPORTUNITY

Sean Spratt still remembers when he heard of the discipline called oenology – that is, the science of wine and winemaking. It was during a mandatory geography course at the Goodman School of Drama in Chicago in 1996: the professor taught geography through the lens of wine. The course fee included tastings of wines from around the world and Sean was captivated. But alas, he was about to graduate in performance theatre and dramatic criticism. He knew he couldn't go back to his parents in San Francisco and tell them he'd chosen the wrong degree. Starting all over again seemed too overwhelming. The moment remains crystal clear to Sean when he said to himself that maybe, in another life, that's what he'd do. But back then he didn't mention it to anyone. Surprisingly, it has turned out to be his parents, Mike and Anne, who helped to make his oenological dream come true.

Sean's grandfather, Frederick Spratt, was a significant colour theory artist, who became Dean of Humanities and Arts at San Jose University, where he worked from 1956 until 1989. He developed a love of wine, amidst the blossoming of the Californian wine industry, which he instilled in Sean, allowing him to select and taste wines from the cellar for family meals. Sean developed a strong bond with his grandfather, and later they crossed paths frequently in the many spaces where art and theatre collide.

Sean's parents met in high school and early on, during their undergraduate studies, Sean was born. After gaining a doctorate, Mike Spratt became a behavioural psychologist, and worked as a mergers and acquisitions consultant. When they were both partners at PricewaterhouseCoopers, he and his friend Mark Feldman wrote an insightful and humorous guide to the world of mergers, acquisitions and corporate change. Five Frogs on a Log was published by Harper Collins in 1998 and went on to be a bestseller.

Anne did a double degree – one in microbiology and the other in chemistry. Her professor for organic chemistry owned a small vineyard and taught his course through the purview of winemaking. It was a smart way to procure willing labour while enchanting the imaginations and senses of his students. Anne loved it but never imagined it would become her future. She worked briefly in the fledgling Biotech industry, moving into the semi-conductor industry in Silicon Valley. Further study led her to become a chip design engineer working amidst the explosive growth of technology during the mid-to-late 90s.

Sean too had spent a few years working in the IT industry to raise the funds he needed to finance his theatre studies. Having been an ace chemistry student in high school, he took to the IT industry like a duck to water, and returned to it once he'd graduated, not wanting to follow the traditional route of turning to hospitality to support his theatrical endeavours. Everyone was talking about this thing called the Internet, and businesses were striving for representation on a new networking device called the "World Wide Web". Sean's network engineering work was exciting yet intense with long hours and very little down time. At the completion of a job, Sean would take off to some far-flung location in Asia, and immerse himself literally in his passion for scuba-diving.

In 1998, Mike and Anne were in need of a long holiday. There was hearty discussion and debate about where to go. Anne wanted to go on a Safari in Africa. Mike's only criteria were to be able to drink water out of a tap and not be eaten by a lion. Mike's Mum had visited New Zealand and adored it, so they spent six weeks travelling the length and breadth of the country. Back home, Sean picked them up from the airport and, in the half-hour ride home, became convinced that his parents would move to New Zealand.

A year later, Sean accompanied his parents on their second trip to New Zealand, just to make sure they were not making a crazy decision. Halfway through the trip, even Sean was wondering how he could find a way to live here. The last stop on their itinerary was Waiheke. Friends of friends picked them up from the ferry and, in a very short space of time, the collective jaws of the Spratt family were hanging open in amazement. It reminded them of rural parts of northern California they loved, with sandy beaches and picturesque vineyards producing their favourite Bordeaux-style wines.

Many lucky and wonderful decisions we make are due to opportune timing. When Mike and Anne began looking for property on Waiheke, stunning coastal parcels were still available and very affordable. They found five lots for sale on the Woodside Bay Estate. Having lived through explosive property inflation in California, they couldn't believe how affordable the lots were, so they bought all five of them. Anne was impressed with the wines they'd tried on the island and, recalling her organic chemistry experience at university, decided that they'd plant some vines in the back garden and make wine in the garage as a hobby. Her hobby was to become a very successful family business.

Mike is a firm believer in knowing your strengths and weaknesses. Even though you're brilliant at one or two things, doesn't mean you will be brilliant in everything you undertake. He made sure that every step of the way into Anne's hobby they sought specialist advice. Through contacts at UC Davis, they engaged Dr David Jordan, a highly respected viticulturist and a familiar face in Waiheke's wine community to assess the possibilities of their land. The site is a natural bowl-shaped amphitheatre with clay soils. It's north-facing and is protected from the cold sou'westerly winds by a ridgeline. There are slopes of 38%, making the land difficult to work, but overall it was a perfect site to grow grapes, especially Cabernet Sauvignon. Unwittingly, the Spratt family had stumbled upon a jackpot in terms of a superb site to grow Bordeaux varieties.

David Jordan's glowing assessment instilled confidence that they could produce world-class wine and their initial decision to plant a couple of acres in vines grew. The project was expanding and they needed help, so Mike discussed the exciting vineyard and winery project with Sean, asking if he would like to become involved. Unexpectedly, a new life as a winemaker was being offered to him by Mike. Sean revealed the secret story of how he'd missed his opportunity to study wine. So many facets in their combined histories were culminating in this new venture: Mike's business acumen and strong friendship with Eric Baugher, the winemaker of their beloved Californian Bordeaux blend, Ridge Monte Bello. Anne's experience in micro-biology and chemistry and interest in the science of wine. Sean's inherited passion for wine, and a natural affinity with chemistry. The family's shared love of cabernet blends.

Sean began talking to contacts at Opus One and Ridge wineries in California, mining them for information, which they were more than happy to share. The first block of Merlot was planted on 4th January 2000, named the millennium block, and they continued planting until 2004. The wine industry in New Zealand had taken off and there were waiting lists for root-stock and clonal selections, delaying planting schedules. In total they have fourteen acres planted with 60% Cabernet Sauvignon, just under 20% Merlot, just under 10% Cabernet Franc and Malbec and 2% Petit Verdot.

Mike and Anne moved here permanently in 2000, and Sean in 2002. He considered doing the Wine Science Programme at the University of Auckland but the task of setting up the business was all-consuming. Mike and Anne studied wine science by correspondence via UC Davis.

Anne went on to complete her Masters of Wine Science at the University of Auckland and they have forged an on-going relationship with the Wine Science programme. They place high value in education and have continually done research projects in conjunction with the University as well as hosting cohorts of students for site visits and tastings.

Their aim to make a world-class wine meant investing heavily in the capital and equipment to eliminate everything in the way of that goal. The winery was built 500 meters away from the vines and they put a bottling line onsite. They built an in-house science unit for Anne's specialised analysis. Luc Desbonnets worked as winemaker from 2004 until 2008, when Sean took over. The Spratts worked very closely together with Luc, and benefitted greatly from his deep commitment to, and understanding of, the wine style and quality they were crafting.

Naming the vineyard was accomplished by telling friends their story. The same word kept surfacing. People spoke of destiny – that everything happened almost without them consciously choosing for it to happen. The Spratt family did everything in reverse. They never imagined themselves planting grapes and making beautiful wines. Glimpses of a possible future flashed at them briefly throughout their lives, but on Waiheke Island, destiny knocked and, in 2000, they opened the door. Destiny Bay produces three Bordeaux blends. Mystae is named after the mystics and followers of Dionysus, the Greek God of wine and fertility and a patron of the Arts. The Magna Praemia benefits from years in the cellar and, as its name translates, you will reap the 'great rewards' if you are patient. The final blend Destinae reflects the unexpected fate this tight-knit family have created. It speaks to all of us who look back in a kind of perplexed astonishment when we find ourselves in a place and time that we did not foresee.

LOUISE THOMAS AND MICHAEL HOLDEN **MICHEAL'S**

—

WINE FLIGHTS

When you live on an island, every time you hear a rescue helicopter fly overhead, you wonder who is on board, and wish them all the luck in the world. Helicopters have delivered both life and near-death to Michael and Louise – and it was flying that brought the couple together in the first place.

Surfing led Michael into flying – although at first it seemed it had just led him into trouble. He spent too much of his second year as an engineering management cadet in Auckland catching waves, resulting in a failed exam paper. The penalty was harsh - he could become an apprentice or leave. He left and joined the Royal New Zealand Air Force, where he trained as a pilot. He spent the next twelve years flying planes and helicopters in military service. The strict routine was an easy adjustment after a lifetime spent at boarding school.

Michael was on an exchange with the Australian Air Force when a helicopter mission nearly killed him. He was piloting an Iroquois during an exercise when a clamp on the tail rotor failed, and the chopper went into mechanical failure. He had to force land it, crashing down through trees in the bush. He managed to get out, only to discover that his co-pilot was trapped and burning. Without hesitation, Michael went back in to free him and incredibly, they both survived.

For this act of bravery, Michael is the only New Zealander to have been awarded the Australian Star of Courage; he is Michael Holden, SC. He received the award in Canberra from the Governor General, and then his colleagues presented him with a celebratory beer mug fashioned from part of the self-same tail rotor that had caused the problem in the first place. Michael keeps the beautiful citation manuscript, the medal and the mug close at hand, constant reminders of the day when luck and life did not desert him.

Meanwhile, Louise completed a degree in Japanese from the University of Auckland and spent three years living in Japan teaching English. In 1993, she returned to Auckland to train as a Japanese teacher, but a friend persuaded her to apply for a job with Air New Zealand instead. The airline was looking for Asian-language-qualified cabin crew and, within a few months, Louise was flying long-haul around the globe. She'd spend a few weeks away, and then return to the little bach she shared with a friend in Palm Beach on Waiheke. She loved the lifestyle and the commute was easy.

Louise and Michael met in Houlihan's Bar in Los Angeles in 1993, across the road from the hotel base for Air New Zealand cabin crew. Michael had quit the Airforce in 1985 for commercial flying and Louise was joining Michael's crew for the flight back to New Zealand. On arrival in Auckland, he gave her a lift to the ferry terminal. A few days later, he came to visit, and they've been together ever since.

Michael was living in Freemans Bay, co-parenting his three children from a previous marriage, but he already had a longstanding connection to Waiheke. In 1970, after completing high school, he had lived with his aunt for a while. She decided to buy a property the bottom end of the island and Michael found her a place in Pearl Bay. Now, he wanted to bring the kids over to Waiheke for weekends and holidays, and decided to buy a place where they could all live.

He loved his time spent out at Pearl Bay, so in December 1994, when the real estate agent told him about a block of land out the bottom end of the island which nobody was interested in, he jumped at the chance. They drove out to the 23 acre block, where Michael climbed a tree to see the views a future house could be built to capture. He was thrilled - coincidentally, the land included a steep north-facing slope – perfect for the vineyard he'd always dreamed of.

From air to earth: Louise was very interested in earth building and booked them into a workshop on the island. The temporary plywood shelter that Michael threw together became a three-bedroom rammed earth house they named The Lodge. Work began in May 1995 when Louise was four months pregnant. She gave up flying, and worked full-time on the house. As soon as the shower was installed, they moved into The Lodge, which was otherwise a shell. Cooking was outdoors on two gas rings during a very wet winter. They were working furiously to enable Louise to give birth at home and the bath was completed the day before she went into labour.

Alas, the wished-for home birth was not to be. Louise's labour was taking too long, so her midwife arranged for a transfer to Auckland Hospital – via helicopter. Michael had built a shelf for the landing, should they need it. The co-ordinates were phoned through to the pilot and they whisked Louise up to the site, clad only in her dressing gown, a drip feed in her arm and there she lay, listening to the chopper hovering nearby. The chopper crew had misinterpreted the co-ordinates Michael had given them, and couldn't find them. It was pandemonium: Michael's 12 year-old son Sam was on the phone to the pilot explaining where they were and a neighbour in a yellow bulldozer arrived on the scene wondering if he could help.

In the end, Louise was bundled into someone's ute and driven to the nearest landing site up the road and within ten minutes they were in hospital. The doctor popped out of the room to retrieve some forceps to assist with the birth – and returned to be greeted by beaming parents and their new daughter Phoebe. After a shower and a cup of tea with friends, the family made their way downtown through the crowded Santa Parade, back to their island home.

Back home, they kept developing their land. They built "The Stable", and ended up living in it when The Lodge was rented out for holiday accommodation. Michael was also considering planting a vineyard – he had been thinking about it for years. He and Louise were friends with David and Veronika Evans-Gander, who also built an earth home, and they watched as their friends planted Passage Rock Vineyard and began making wine. They developed another friendship with Tony and Moira Forsyth who shared the trials and tribulations of owning Te Whau Vineyard and operating the restaurant.

Tony introduced Michael to Herb Friedli, their winemaker, who agreed to have a look at his vineyard site. He liked what he saw and, knowing they loved red wine, Herb suggested they plant Syrah. He also suggested very close planting on their small and steep site. They chose an ancient European style of planting, named Goblet, also known as "bush vines" with no poles, wires or trellising. Two metre high stakes were driven into the clay, one metre apart, along rows that are one metre wide. Each vine is tied to a stake and every single job, by necessity, is done by hand, including mixing and spraying the organic preparations. The first 100 vines were planted in 2001, and planting continued each winter for three years. Their first harvest in 2005 produced just 67 bottles of the 'Michael Syrah'.

During those early years of planting, Louise and Michael were also building a large rammed and poured eco-friendly earth family home, together with six fulltime workers. (The day of the first vineyard planting, Louise was heavily pregnant with their second child Alice, and Michael was away flying, so Louise commandeered the builders to help out with the painstaking work). Two of the workers lived with the family which now included Phoebe, baby Alice and Michael's three older kids, who would come and go. The proposed one-year build dragged out into three years and every possible revenue stream was poured into the project. Once again, they moved into the unfinished house to free the Lodge and the Stable up for holiday rentals and the occasional wedding. Their son William was born in the bath outside in 2005, with Michael and the mid-wife arriving just in time for the birth.

But then, Michael and Louise ran out of funding. They had no choice but to sub-divide the property – one section with the Lodge and the Stable on it and the other with the newly completed home – and put both properties up for sale. They would have to live in whichever did not sell first. Survey plans ensured the boundaries could change to include the vineyard on whichever property they were left with. The real estate market was slow, but long-term rentals generated an income.

Entering the final home that they built, you melt in its air of tranquillity. You could spend days gazing out over the infinity swimming pool as it disappears into the ever-changing colours of the ocean reaching out to the Coromandel Peninsula. Yes – they were able to keep this family home. Friends bought the Stable and Lodge property and, soon after, Michael and Louise rented their home to an American family for 18 months. They rented a house just minutes from the school, needing to make sure that life at the bottom end of the island, with two hours driving to and from school daily, was still the right choice. And it certainly was and still is, although they consider themselves fortunate due to good luck rather than good management.

Their wine production has grown to 600 – 1000 bottles annually, depending on the vintage variables. Herb has remained consultant, but after years of instruction and training, Louise now manages the vineyard and Michael, who has retired, works alongside her. The fruit is picked and transferred directly to Te Whau Vineyard, where both winemaking and elevage proceed under the careful guidance of Herb and Curtis Staples. The wine is listed at a few notable restaurants and private sales often arise from restaurant diners.

The Thomas-Holden family are flourishing, like their vines; and they are savouring the fruits of their labours.

LANCE AND LUCIANA BLUMHARDT
JURASSIC RIDGE

—

WINE AND GEOLOGY

Lance is an extraordinary man with an extraordinary mind. Get him talking about the name of his label, Jurassic Ridge, and you could spend days acquiring a new love and knowledge of geology. If, he explains during our interview outside the cellar door, we'd been sitting here during the Jurassic period some 155 million years ago, we'd be 10,000 metres high on a mountain. Imagine, he continues, something like Mt Cook here on Waiheke, just 12 kilometres from Auckland's CBD. "It would be absolutely fantastic – we'd be skiing and swimming on the same day".

Millions of years of erosion exposed the core of the Jurassic mountain smack-bang in the middle of his vineyard. Lance continues to set my brain alight as he tells me that just by walking from Fossil Bay or Tairere Cove, about half a kilometre from where we sit in the soft morning light, we are travelling across 130 million years in geological time. And it turns out that geology is only one of many sciences that Lance has delved into.

Lance speaks with a fine British accent and I'd assumed that he was English. It was a surprise to hear that he had was born in Nelson and had been coming to Waiheke since he was 8 years old. Lance's uncle Stewart had a boat and they would sail around the island, anchoring in beautiful bays where they'd fish for dinner, fire up the billy and sleep on the beach. They'd stop in at the store in Orapiu to buy milk, bread and the daily newspaper – that's about all they could acquire. This was Waiheke in the 50s and they hardly saw anyone. Eventually his uncle moved to Waiheke, so they came to know the island well.

After high school, Lance first studied geology then medicine, before graduating in neurology and training at Wellington Hospital. He went to Europe to research for a higher degree and stayed for 30 years, becoming a professor of neurology and consultant neurologist at Queen's Medical Centre, at Nottingham University Hospital.

He met his wife through his profession: Luciana is a practising neurophysiologist, who has worked in New York and travelled the world with her research. She was born in Taurasi, Italy, home to olives and the famous Aglianico grape, and Italy is where the couple met, in Sorrento, 1987, when Luciana was delivering a paper at a conference session that Lance was chairing. As fate would have it, they bumped into each other a few days later in the romantic coastal village of Positano and their relationship bloomed. They married and lived in England, and have two children, Hannah and Luca.

In 1992, the family travelled throughout New Zealand's wine regions. Lance was keen to relocate permanently and had convinced Luciana to leave family and friends by surrounding her with grapes and olives of their own. But they couldn't find a place to call home.

Just before returning to the UK, the couple visited Lance's Uncle Stewart in Church Bay. They spoke of their plans to plant olives and grapes – of their unsuccessful attempts to find somewhere to live. I imagine that Stewart roared with laughter as he told them that what they were searching for was right under their noses. All they had to do was to walk up the road to a new sub-division of the Church Bay Farm. That afternoon they found a block named Church Bay Rise that was perfect: gentle slopes including seven acres of forested reserve on the south side, providing shelter and protection from the prevailing cool winds. It was just the right size for a one-man band. The geological puzzle was waiting to be discovered.

Lance sat bargaining over the sale price while Luciana and their two babies were safely en route to the airport – he almost missed the boat and the flight. As their plane left Auckland, it flew right over the island and they could look down on their new home – "it felt really great, albeit a bit nerve-wracking" remembers Lance.

During the following decade in Europe, the family often holidayed in coastal Bordeaux, learning about the great wines of the area. They also holidayed regularly with Luciana's family in Italy, where they bought a house. But Lance felt his kids were becoming British, and he wanted them to experience life in New Zealand. They decided to relocate to Waiheke earlier than planned, with Lance swapping neurology for oenology.

As a neurologist, he was fully aware that upon 'retiring', it's not a good idea to sit around on the couch and atrophy. But he knew that the vineyard wouldn't let him: Kim Goldwater had once told him, during a Waiheke holiday grape harvest, that there was enough to experience in viticulture to last three or four lifetimes.

They had been visiting the island every two years to plant and prune olive trees and 1000 native trees along the boundary

– introducing their kids to island life. In 2001 they moved permanently to Waiheke. Ever the scientist, Lance had collected weather data on his land from the outset in order to match his grape varieties to the data. He planted Cabernet Franc and the Italian varieties: Montepulciano, which loves its maritime climate in Abruzzo and Fiano d'Avelino, a beautiful white variety that grows in the vineyards surrounding their Italian home. The main blocks are planted in Syrah and he also has a small amount of Pinot Gris. He augmented his tiny crop by buying a vineyard in Marlborough's Waihopai Valley, planted in Sauvignon Blanc.

After helping to plant the 1000 vines, Luciana had done all the work she wanted to do on the vineyard. She wasn't ready to retire and so she holds neurophysiology clinics all over the country, including Waiheke, and does research at Auckland and Tauranga hospitals. Her busy schedule means Lance has had to run the vineyard by himself. Well – it was his retirement plan after all.

They finished planting in 2002 and Lance knew he had three years to think about wine before his first vintage. He'd read that one cannot be a winemaker and a viticulturist simultaneously, but decided to challenge himself. He completed a diploma in winemaking by correspondence, occasionally travelling to Hawkes Bay to do work in the labs on site. He allows that there are real conflicts when you have to meet deadlines in both vineyard and winery. "It can get a bit desperate at times, but it's fun trying," he says. This is Lance all over. He is fluid and calm, yet you just know that his brain is cross-examining everything. He's a wonderfully wise man, quiet and serious, yet quick to laugh at himself and the situations he's bumped into.

This geologist really knows his soil and his organic focus is to nurture the land, leaving it in a better state than when they first acquired it. He's become accustomed to the back-breaking task of mowing under the vines with a weed-eater. It takes him four days, but it's a task required only four times each year. In between the rows the ground is left alone, sprouting natural pasture with lots of wild flowers at different stages of the year, sometimes growing to waist height. During the heat of the summer the pasture dies off, but the clay retains enough moisture to keep the vines thriving. He'd love to go solar; the panels on the roof lie ready, he's just waiting for the battery technology necessary to operate his home and business.

During the very demanding 2011 vintage, he started to get worried. "The fruiting season poured with rain and it was utterly dreadful," he remembers. It was the one and only time he'd ever seen botrytis – the fungal infection on grapes – in the vineyard. I can hear the collective sighs and groans from every vineyard owner, winemaker and viticulturist on the island remembering that season. But Lance got lucky. In the door walked Danny Schuster, an internationally renowned wine consultant with vast experience who works with the famous producers Ornellaia in Italy and Stag's Leap in the Napa Valley. Lance didn't recognise him at the time and says that Danny simply asked if he would like some help to get the most out of the vineyard. "We've had a great thing going ever since – he's been fantastic," Lance enthuses. Danny has tweaked the pruning and has made changes in the winery as well.

In 2015, Lance employed Adriano Dumitru, who, like Danny, just turned up one day. Lance knew of his work in Italy, as head of the viticultural team at La Sorvana, a beautiful estate in Tuscany that he and Luciana had visited several times. Lance just couldn't believe it and is delighted with his presence. He can finally allow himself time away from the vineyard to go back to Italy with Luciana each year for more than two weeks at a time, and to take his boat out sailing here on the Hauraki Gulf.

His cellar door has become wildly successful, which is very handy because the one thing that Lance does not like to do is to market his wine. A woman from Finland came to the cellar door and bought a few cases of Montepulciano to be shipped back home. She became a repeat buyer and, after she took his wine to parties in Helsinki, word of mouth spread - Finland is now Lance's largest export market. About 85% of his visitors are international, most revealing that they've never tasted wine quite like his. He's not sure if it's the winemaking – his 'not-mucked-about-with' philosophy, using whole berry ferments, allowing longer and slower ferments – or if it's the cellar aging of the wines so that they are at least four years old upon release, ensuring the tannins are beautifully softened, eliminating the need for fining. Or maybe it's his organic vineyard practice. It is, of course, a combination of all of these things, combined with a perfectly sited vineyard to ripen his well-chosen fruit.

And that Jurassic rock can't hurt.

CHRIS CANNING

—

CHRIS CANNING
HAY PADDOCK

—

WINE AND THE CHAMELEON

Hay Paddock vineyard was hatched out of a rather casual conversation Chris Canning had with his neighbours on Waiheke, Bryan and Heather Mogridge. They had sons of the same age who had not yet made career choices. Bryan and Chris had extensive experience in the wine industry, so when land came up for sale just down the road from their Onetangi holiday homes, they decided that it would be a great idea to plant a vineyard, make the very best red wine they could, and get their sons to manage the operation. Chris, who had owned a vineyard and was a winemaker in a previous life, would oversee everything. Bryan's role was purely as investor. And the boys – well, Waiheke is not really that appealing to young men in search of a career. So that left Chris to get out into the vineyard and do it himself.

Chris is definitely one to put in the hours for whatever project he's working on. But in 2003, when the vineyard was planted, Chris was 62 years old. He imagined that he'd be spending time pursuing his love of writing, not planting vines and making wine all over again. But the other thing about Chris is that he doesn't sit around feeling sorry for himself – he just gets on with it. He's incredibly adaptable – as you will read in this story of the many lives of Mr Canning.

As a young cadet journalist, he took off from Auckland for the bright lights of Sydney and then London, where he lived for 17 years. He was 21 and travelling through Europe when a friend suggested he join him in Burgundy harvesting grapes, working alongside a bunch of gorgeous girls. He stayed on, not for the girls, but to feed his growing interest in what was happening in the cellar. He dipped not only his toes into the world of winemaking, but his entire naked body into the wine vats at night, where, by the light of a candle, they had to manually stir the grape skins through the fermenting wine. This was 1962 in Burgundy and the modern age of winemaking was light years away.

Back in London, he worked in public relations and advertising, eventually setting up his own small advertising agency. Thanks to arcane financial controls, Chris' company ended up lending money to a client who wanted to buy a slice of paradise in Tuscany: 135 acres of vineyards and olives. Chris spent summer holidays at the Tuscan dream farm, and travelled down every October for vintage, working hands-on with picking and plunging.

He and his business partner opened an office in New York, so Chris took off to conquer another of the world's great metropolises. He was commuting from New York to London every week (a bit more high-pressure than the Waiheke-Auckland daily floating social networking waterside café/ wine bar commute!). For three years he persisted with the tiring schedule until he realised the ill effects his lifestyle was having on his health. When a tape recording arrived in New York from his 5 and 6 year-old children, he knew he needed to make drastic changes: he couldn't believe that the perfect little English voices reciting the nursery rhymes were his own children. He decided to sell his share in the business and find a place in the world to accommodate his family, preferably where he could grow grapes and make wine.

In the late 1970s, the Californian wine industry was in great shape. The Judgement of Paris in 1976 resulted in Californian Wines winning Gold medals for a Chardonnay and a Bordeaux blend, snatching them from their world-famous competitors in Burgundy and Bordeaux in France. And in 1978, a young wine writer named Robert Parker began to critique and acclaim his local Californian wines. Chris researched buying land there, only to realise the exorbitant capital gains taxes he would have to pay if he stayed in America. After many years abroad, the family returned for a classic Kiwi summer holiday: sailing, swimming and barbeque feasts. Their experience convinced them it was time to return to New Zealand.

In 1976, the Canning family bought a farm in Mangatawhiri, South Auckland. As usual, Chris had plunged headlong into a complex situation and had a hell of a lot of adapting to do. His life changed overnight, from donning a double-breasted suit, to buying 1000 sheep and gumboots. He literally had to teach himself how to ride a horse and work with a sheep dog. He had a farm manager, so he wasn't flying completely blind. The kids loved it and life was good.

And he finally got to plant his vineyard: Chardonnay, semillon, Cabernet Sauvignon and Merlot. He tended the vines, introducing himself to the West Auckland wine community, spending as much time as he could on vineyards and in wineries, gaining experience and understanding.

In 1980, Chris harvested his grapes and crafted his first wines. He opened the doors of the deRedcliffe Vineyard tasting room. It was one of the only areas outside of West Auckland where you could go and taste wine at the Cellar Door. One of his visitors was a knowledgeable wine enthusiast called Kim Goldwater. Another, was an incredibly keen young man who was absolutely bitten by the wine bug called Stephen White. Little did they know they would all end up planting vineyards on Waiheke.

That same year, deRedcliffe Estate joined the New Zealand Wine Institute and was amongst its first 20 members. The Institute represented the growing needs of grape growers and wine makers in every aspect possible – from drafting legislation and regulating the industry to marketing for international sales. Chris attended meetings alongside my father Terry Dunleavy who was CEO and Bryan Mogridge, then-Chairman of the Board, who would become his future neighbour and business partner. By 2015 membership totalled 673 – that's how fast our wine industry has grown in 35 years.

Demand was high for Chris' wines, which sold out quickly, as there were very few producers of Bordeaux blends in Auckland at that time. But he recognised that his location was not the best for growing red grapes, so to increase volume and satisfy both demand and his own quest for quality, he started buying in fruit from Hawkes Bay and Blenheim. The fruit was sent up overnight in gas-insulated grain trucks. His marketing background convinced him of the tourism opportunities of his vineyard's location, close to Auckland Airport. Building got under way to create accommodation and a restaurant alongside the Cellar Door and the rebranded Hotel du Vin opened its doors. The timing could not have been worse. The hotel opened on Black Monday, October 19th – the day the world share market crashed.

By now you understand how adaptable Chris is. He knew he had to make the best of his bad situation. Once again his marketing skills came to the fore. He invited domestic and international travel journalists to stay and indulge in everything on offer at Hotel du Vin. One of those journalists went back to Japan and described a paradise discovered, headlining the front page of Tokyo's financial newspaper. An investor in resorts and theme parks read it and contacted Chris to see if he was interested in selling. With impoverished wallets snapping shut across the globe, he said yes.

Chris then moved to Sydney and went into business in the property market. In 2001, in search of a holiday destination in New Zealand, he remembered time spent sailing around the Hauraki Gulf. He rented a bach on

Waiheke Island, and even though it rained for the entire two weeks, it left a lasting impression. Chris bought land on the Onetangi clifftops for future holidays with his extended family. His neighbours were also building next door and by incredible coincidence, those neighbours included Bryan Mogridge, whom he'd met at the Wine Institute all those years ago. And so began the conversation that sparked the creation of Hay Paddock vineyard.

Chris had tasted David Evans Passage Rock Syrah and loved it. It encouraged him to plant mostly Syrah, along with a small amount of Petit Verdot for blending purposes. In 2006, their first harvest off three-year-old vines produced a small vintage. From the outset, they had decided to bottle-age the wine for two years before release. Not everyone has the luxury to do this as most vineyards require the capital from early wine releases to plough straight back into their cash-hungry business. The 2006 vintage won Trophy for Shiraz/Syrah at the Hong Kong International Wine and Spirits Competition, setting high standards for future vintages. Hay Paddock wines are made at Mudbrick's Shepherd's Point winery in Onetangi. And now – as Chris is in his seventies – Mudbrick Vineyard manages the Hay Paddock vineyard, and Chris is consultant winemaker.

Chris and his partner Jules now live on Auckland's North Shore where she continues to manage sales and distribution. Hay Paddock Cellar Door operates out of Rangihoua Estate, which produces award-winning olive oils in Rocky Bay, founded by Colin Sayles and Anne Stanimiroff in 1996. By age 75, Chris was working on a screen adaptation of his second novel "Songs for Leonard" for an interested party in the LA film business and, separately, had won a Hollywood screenwriting competition, necessitating fulltime work on another screenplay. Out of 6000 entrants, all of whom were given the same synopsis, they selected Chris' entry. Once again he took off down a road he'd never travelled, this one taking him to Hollywood. He's also written and recorded the songs that haunt his novel, together with Jules, and Eden Roberts, a Kiwi singer-songwriter living in Los Angeles.

Chris Canning is finally spending his days dedicated to his passion of writing. As he remarks "I used to be a winemaker who did a bit of writing and now I'm a writer who does a bit of winemaking". But who knows? Another surprise could be just around the corner for the ever-adaptable Mr Canning.

DAVID AND HELÉN GUNSON

DAVID AND HELÉN GUNSON
WOODSIDE HILL

–

WINE: AN OUTDOOR PURSUIT

Summer in Sweden in the late 1980s was idyllic for David and Helén Gunson: the IT consultants bought a boat and spent holidays and weekends sailing the Stockholm Archipelago's 24,000 islands, in the long summer light. But then there's the Scandinavian winter… and the couple's desire to get out of the office all year round grew alongside their passion for wine. It's been a long journey from Swedish IT to Waiheke viticulture, but now they're wonderfully settled on Waiheke, perched on their vineyard above Woodside Bay.

When David finished high school in Palmerston North in the 1970s, he tried to join the Navy. But just before his cadetship to become an officer began, the Rear Admiral himself called: with deep regret, they had to turn his application down. David was too tall. Critical injuries had been sustained during an on-board accident because the cadet was too tall. They could take no further risks and had to implement height restrictions.

A change of plan was urgently required as the university year was about to begin. David's dad was a lawyer and his Mum was a Professor of Information Technology. David, typical teenager, didn't want to follow in his parents' footsteps, but he had inherited a mathematical mind, so chose a degree in Computer Science. It would enable him to work anywhere in the world.

Helén was born in Stockholm, Sweden. Her father invented extremely high precision metal tools, including a piece of equipment to balance turbine blades in jet motors. Her mother was an x-ray nurse. Helén had a keen interest in biology and the medicinal properties of plants. Alas, she developed a debilitating allergy to the laboratory environment and so she moved into the fledgling world of IT because she too had a natural affinity for mathematics.

Stockholm is a city full of illustrious companies – Ericsson, Ikea, H&M, Spotify, Electrolux and Volvo – and is second only to Silicon Valley in terms of tech business proliferation. The IT industry expanded quickly in the eighties and its bright young stars – including Helén – lived life in the fast lane.

David finished his degree and took a job with NZ Post, precursor to Telecom New Zealand. He was part of the IT department working on upgrading the entire phone network and, to his surprise, was selected to travel to Stockholm to work with the vendor to design the programmes to enable the upgrade. The project stretched from six weeks to a year, with frequent travel. Helén was working on the project in Stockholm and their friendship blossomed. She was sent to Palmerston North after David's return home, to support the project installation. During her six-month residency, their friendship became a romance, and in 1987, David moved to Stockholm.

They continued working in IT – the working culture in Stockholm is inspired. Each company has particular interests and employees select their workplaces where the hobbies match their own. Helén's company pursued skiing and sailing – both interests that David shared too. While working at Ernst and Young in Stockholm, David and two of his colleagues formed a company selling software development tools. They soon realised that what the market needed was training and support, and so developed a novel way of understanding how to use software tools and how IT departments could function. It was a great success, and over seven years the company grew to employ eighty staff.

Outside of work, wine – along with sailing – became a central focus for David and Helén. All wine for sale in Sweden is controlled by a state-owned monopoly called the Systembolaget. The monopoly is open to suppliers large and small throughout the world, with quality being the major criterion for selection. Sweden is famously neutral in its judgement of wines, believing there is no reason to elevate wines that do not merit a great review, no matter how famous they are.

In Sweden, the couple's access to good and very grand wines of the world was fantastically broad. David's business partners also had a passion for wine and they held tastings in the cellar of a 13th century building housing their wine collections. Slowly David and Helén's palates grew to distinguish the many differing characters and nuances of wines from around the world. Years later, when they returned to New Zealand to live, they were invited to an informal tasting of international wines at Kennedy Point vineyard, an event with a convivial yet competitive edge. David scored highest and Helén came in second, not, they say, because they had the best wine palates but because they had such extensive exposure to the wines of the world in Sweden.

In 1992 the couple came back to New Zealand for David's brother's wedding and toured around the country with family and friends. Post-holiday, back home in Sweden, David developed a bad allergy to the birch trees that flourish everywhere. Heavy medication was the only way to control the allergy, but it made him miserable, tired and moody. The long dark winters were spent indoors, alleviating his symptoms, but the arrival of spring and summer created havoc. Discussions about moving to New Zealand began.

Balancing climate extremes meant life in Sweden was spent mainly indoors. They were both adamant that if they moved to New Zealand, they'd spend as much time as possible outdoors. Their love of wine had developed into a passionate affair, so unquestionably, that would become their future. This couple belong to the small but very particular clan for whom wine gets under the skin. This clan seems unable to eliminate their intention of buying land to plant vines and make wine. They are all slightly mad and brave and they keenly dig holes and plant never-ending amounts of money and sanity inside.

A New Zealand holiday was planned in 1997 to investigate wine regions that were a modest commute to a city. David knew he'd have to keep working in IT. Helén would finally get to use her long-held love of plant biology and manage the vineyard. They looked at land in Waipara, close enough to commute to Christchurch, and in Matakana, close to Auckland, but it was the beauty of Waiheke that impressed itself upon them.

They spent a day driving around the island with the Rural Postal delivery van. With commentary from the postman, they went everywhere, delivering mail to private properties including those not easily reached. It gave them a great sense of the scale and contours of the island. Then on a whim, they walked into a real estate office and were taken to view several properties. It was a typical November day, with ever-present spring winds tearing across the landscape. At each location the wind snatched at their clothes and slapped their faces.

The last property of the day was on a new sub-division in Woodside Bay. They wound their way around Whakanewha Reserve, eventually scrambling up a steep rutted gravel road to what felt like the top of the world. Panoramic coastal views surrounded a large flat area that gave way to slopes of 20 degrees facing north-east through north to north-west. They wandered down the slopes and into a wind-still landscape. Protection from prevailing wind is a great asset for grape growing and they were excited. They arranged for further professional analysis.

David's father was a property lawyer and knew all the right people to investigate the pros and cons for them. Dr David

Jordan did the necessary analysis of the site and the soil, reassuring them that it was perfectly suitable for planting vines. An engineer assessed the layout of the land and the practicality of working the site. As long as they invested in a superhero tractor, everything was fit for purpose. The deal was done – they signed and sealed their future and went back to Sweden to earn as much money as they could to help finance their new lives.

It wasn't until 2004 that they were able to move to Waiheke Island. Then work got underway: posts were punched down the slopes and a German tractor was ordered to navigate the 1.8 metre wide and very steep rows. The house build was slowly progressing. Helén worked with Doug and Anne Hamilton from Peninsula Estate (no longer in production), gaining valuable experience with harvest, pruning and canopy management. While there, she met Mike Wood who was working at Passage Rock. This trio cemented a friendship and Mike agreed to work with them as winemaker to create Woodside Bay wines. He taught Helén and David a tremendous amount, both in the vineyard and the winery, and remains their consultant winemaker.

In November 2005, they planted 3000 vines that march down the slopes on one acre, just a few short steps from the house. Distressing building delays were softened only by their speedy acquisition of superior clones and rootstocks. A great deal of research continues to elevate the quality of plants available, enhancing outcomes on the vineyard and in the bottle. Helén and David's mandate was to make a wine they loved and would enjoy drinking throughout their lives. They chose to craft a wine in the style of Saint-Émilion in Bordeaux, a blend that is typically Merlot and Cabernet Franc dominant. Their wine 'Emile' is a blend of 70% Merlot, 25% Cabernet Franc and a tiny amount of Petit Verdot. In a great vintage, they barrel-select a blend for their Reserve label. They also have three rows of Viognier from which they produce a sweet white port.

David works weekends in the vineyard and cellar door, and continues his ferry commute to the contrasting world of IT in Auckland. Helén manages the vineyard with her audio-books plugged in. Svea, the lovely black canine in residence, lopes across their kingdom, daring the wild pukekos to step inside the boundary. David and Helén have realised their outdoor life, embracing the changing seasons on their vineyard, overlooking their antipodean archipelago.

Street number or rural route
City or post office
AGE
SEX
SIGNATURE
WARNING

No. 3
IN FULL
Signature of issuing officer)

LUC DESBONNETS
FRENCHMANS HILL ESTATE

—

LA VIE ET LE VIN

Luc Desbonnets – winemaker, vineyard builder and businessman – is a deeply philosophical man with generous insight. He understands that success takes time, experience, courage and belief. This vigneron has consulted for many people over the years, sharing his vinous experience and acute business sense. He's proud to be part of our industry during this exciting time in our history. Excitingly, for Waiheke wine overall, he says: the best is yet to come.

Luc's father, a wool buyer and exporter, migrated to New Zealand from France at the age of just 23. So, on completing a business degree in 1989, Luc took time out to meet his family in France, to learn the language and to decide upon a future for himself. Although the family had no background in the wine industry, his father suggested that while Luc was in France, he should take a look into France's culture and tradition of wine. Luc's interest was piqued, so he went to work on a vineyard in Chinon in the Loire, home to Cabernet Franc and Sauvignon Blanc, to see if the industry was the right fit for him. It was.

Luc studied in Tours, before moving down south to work on vineyards in the Rhone Valley, the Savoire region and Burgundy. He went on to study at the Centre de Formation Professionelle de Produits Agricole a Blanquefort, Bordeaux. It had its own vineyard and winery, Chateau Dillon, located in the Haut Medoc appellation of Bordeaux, very close to the famous Chateau Margaux. From 80 applications, only 18 students were selected, aged between 23 and 58.

A particular memory for Luc is of a class trip to taste a Saint-Émilion Grand Cru, Tertre Roteboeuf, made by Luc's classmate Louis Mitjavile's father, Francois. Luc had tasted many great wines while studying, but in tasting this particular wine he realised that one day he wanted to make wine that was its equal.

On the completion of his studies in Viticulture and Oenology in Bordeaux, he was hired as a flying winemaker for a large French wine company, working for them in Hungary and Moldova – fascinating places drenched in culture and tradition. But with two degrees, diverse experience and growing confidence in his profession, he was ready to come home to New Zealand.

Six years of living, studying and working in France had given him the opportunity to witness the philosophy of people who dedicate their lives to making great wines. Like any undertaking that involves intensive learning, he only slowly began to comprehend his experiences. He didn't realise how much he knew and remains humble regarding his experience and achievements. There is much to be discovered, and in conversation with him, especially over a glass of wine, hours may slip away.

Luc's had some great mentors, too. Back in New Zealand on holiday, his studies required that he complete trainee periods on vineyards. His family lived near Martinborough, amongst inspirational and legendary wine people. He got a job with Larry McKenna at Escarpment Vineyard, working alongside Nick Hoskins, a talented viticulturist in the area. He also met Neil McCallum from Dry River.

Luc returned permanently to New Zealand to take up his first job on Waiheke Island at Man O' War Vineyard, where he worked alongside Matt Allan for four years. They established the winery and developed many small-parcel vineyards, planting high density as Luc had learnt to do in France. When you plant vines very closely together they fight for the right to exist on the infertile soils. They dig down deep into the earth to extract what they need and they retain more heat for ripening. Luc wanted to be able to express the uniqueness of every small parcel of vines he worked with.

Luc's next job was as winemaker for Destiny Bay Vineyard on Waiheke, producing high-quality luxury Bordeaux blends. During his time there, Robert Parker's representative Neil Martin was in New Zealand tasting wines. Neil told Luc that his wine reminded him of a particular chateau in Saint-Émilion. It was Tertre Roteboeuf, the very same chateau that had sparked a desire in Luc to make great wine all those years ago. He felt he had finally proven himself.

During this period he also began making vineyards for people who wanted their land developed and vineyards built. While travelling to and fro from his Rocky Bay home to work each day, he drove past a block of land that kept calling his name. After 10 years in the industry, he'd told himself that it would be crazy to plant his own vines, but after 15 years he was reconsidering his options. One part of his business brain told him it would be too expensive and he'd never survive financially. The other part kept telling him that it was possible, but only if he found a north-facing piece of land above the water with shelter. Now he had found that piece of land and was ready to take another brave leap into his future.

In 2005, together with a new business partner, he purchased that block – his home vineyard, Frenchman's Hill in Rocky Bay. A vineyard is a long-term investment in which you need to look 20-30 years into the future. Luc talks about planting wines not vines, requiring a lot of hard work, authenticity, expression and investment. He believes that if you apply your skill to give the land more definition and permanence, you allow others to carry your vision into their future.

Luc established three estates, all 100% coastal in 2006 which he continues to manage: Frenchmans Hill Estate (FHE) - his family vineyard at Rocky Bay North; Expatrius Estate on the Omaru Bay headland in Orapiu and Te Rere Cove Estate in Church Bay – home to his winery.

At Frenchman's Hill, situated on slopes rising steeply from the inlet of Putiki Bay, eight varieties are planted. Unusually, he blends Tannat and Kolor into the 'Expatrius Blend of 8' label. These varieties are part of the spice wrapped around the core of the traditional Bordeaux varieties of Cabernet Sauvignon, Merlot, Cabernet Franc, Malbec and the lesser-known Petit Verdot. The blend is a reflection of the flavours of Bordeaux and the south of France, where he's made a lot of wine. FHE also produces a Syrah and a Rosé.

Te Rere Cove produces a Syrah, Chardonnay and Pinot Gris. Omaru Bay produces 'Expatrius Syrah' and 'Expatrius Advenus', an atypical blend of Sauvignon Blanc and Viognier, very French in style. In total there are 5 hectares of vines over 3 sites to care for.

To be 'in expatrius' is to live away from your homeland, while remaining very attached to it. The name celebrates migrants the world over who have carried their culture and traditions far and wide to start a new life. Many of these migrants carried their vinous heritage with them to recreate it in a new land and these wines honour those people.

Luc's wife Anna is an interior designer, crafter and artist. Together they operate the cellar door and a restaurant offering a French-style Menu du Jour, while juggling a busy family life. The tasting room houses a shop packed with gifts and homeware offerings.

Luc's business philosophy stretches out into our community. He has an apprenticeship programme for local boys keen for work and who don't relate well to school, preferring to get their hands dirty and their bodies strong by working on the land. He's keen to share his knowledge to enable others positive change in their lives. Luc believes that as an entrepreneur, motivating young minds is a truly great reward and the outcomes flow into the community.

He speaks of the local legends around him, who have been producing wine here for over 20 years. When he arrived in 2000, the reputation of Waiheke was mixed. But now, says Luc, we have some great technicians producing superb quality wines. Over time winemakers get attached to their site, to their specific climates and topographies – the meso-climates and their understanding of what is possible grows. Sometimes change is necessary in order to grow the most successful varieties to suit each site. Waiheke has an extraordinary amount of these meso-climates, which culminates in a kalaidescope of expression.

There is always room to improve, but Luc thinks Waiheke should be able to achieve much in the future, producing high quality wines, commanding good prices and bringing success to the domestic industry and to our local wine community. The outlook, he feels, is very good. Waiheke has attracted many highly skilled and motivated people like the Desbonnets, all with the aim of building a solid foundation for their families and for our community.

PS 11/1

CONNIE FESTA AND ROB MEREDITH
PEACOCK SKY

—

WINE AND PATIENCE

Connie Festa and Rob Meredith worked their way around the globe for decades, but when you visit them at Peacock Sky – their Onetangi vineyard, tasting room and restaurant – you can't imagine there is any other place in the world that they would rather be. In a house perched high above the sea, with a 360° view, this enterprising couple and their lovely dog Sky have finally found home.

Rob is from London and Connie is from Montreal. Connie's Italian father was a farmer who had dreams of his own vineyard; her mother loved cooking and had managed two restaurants during her career. As a young girl, Connie wanted to be an artist, but a successful business career was the only acceptable option as far as her parents were concerned. Connie dutifully set out and built one in the technology industries, and met Rob when they were working for the same company in Toronto in 1987. They soon moved to Phoenix, Arizona in order for Connie to set up a division of the company there. Entrepreneurial Rob – who had established a few of his own businesses before he met Connie – set up another business of his own.

Connie sadly lost her parents early into her business career. After years of sacrificing her life to the business world, she decided to jump into the creativity of the kitchen and spent two-and-a-half years as head chef and manager of her own restaurant in Phoenix. Basically, she swapped long days in the office for even longer days in the restaurant. Lack of sleep dogged her with 5am deliveries and late night ordering and administration. It was rewarding and gratifying because it was her great passion – but she told herself that she would never do it again.

Sixteen years rolled by and, by 2004, they began thinking about a new chapter in their lives. A Kiwi friend whom they adored spoke so passionately about New Zealand that they had to visit and see it for themselves. Lord of the Rings was everywhere and this part of the world was a distant enchanting possibility.

They arrived in Auckland, seduced by the city of sails. A day trip to Waiheke Island confirmed everything they had been told – just 35 minutes from Auckland lay paradise. On they travelled throughout the North and South Islands, but the magic of Waiheke held them under its spell.

At Auckland Airport, feeling despondent that their grand adventure was about to end and downhearted at the thought of dragging themselves through passport control and flying back toward everyday realities, they made a life-changing decision. Before boarding the plane, they purchased tickets to return to New Zealand just three months later. Connie and Rob did not have a plan for what they would do once they got here, but they were absolutely clear that they wanted to live in New Zealand.

Thinking it through, the couple decided to realise Connie's father's dream of owning and operating a vineyard. They made a plan and visited New Zealand regularly, travelling through our wine regions, researching all options. They spent two years trying to close a deal on a property in Matakana, which finally fell through. It seemed that every time they found the right place to buy they missed out on it, sometimes by just 24 hours. They started to wonder if they'd ever find what they were looking for.

Rob and Connie planned to skip the first 4-5 years of planting and developing a vineyard by buying one that was already producing wine. They had no luck, so considered buying land and planting it out themselves, but for this business-savvy couple, the numbers didn't add up. They eventually bought an apartment in Auckland, losing sight of their vineyard dreams. While closing the apartment deal, Rob and his real estate agent had drinks with another agent, who listened to Rob's tales of his and Connie's frustrating and unsuccessful five-year roller coaster ride of nearly-but-not-quite finalising deals on vineyards. The agent told Rob about a vineyard he had listed for sale on Waiheke Island. It was the night before Rob was due to fly back to Arizona. He phoned Connie who told him to stay put and check it out.

Rob viewed the property the next day, but unfortunately it wasn't what they wanted. He scoped another property that same day – but that wasn't appropriate either. Surprisingly, he learned of yet another vineyard for sale and went to check it out on the off-chance that maybe, just maybe, it could be the one. The owners were Sally and John Ewer and the vineyard was Peacock Ridge, where they made wine and sold extra grapes to other vineyards. Connie and Rob were looking for a business with a base to build upon and an appropriate size for them to manage. This vineyard had been planted in 1999 – three hectares of mainly Cabernet Sauvignon and Merlot, with small amounts of Cabernet Franc and Malbec. The vineyard had been meticulously laid out using GPS and laser alignment, John's engineering background clearly evident. The first wines had been produced in 2002 and the Reserve Merlot won a Silver medal at the Air New Zealand Wine Awards.

Everything looked promising but Connie and Rob's scrutinising business minds vetted every imaginable aspect of the business. They hired the viticultural expertise of Allan Clarke to assess the vineyard. They took wines produced over the years to Kim and Jeanette Goldwater's home to be tasted blind, eager for experienced comments. On the plus side, as the vineyard had very limited prior marketing and exposure, they could rebrand and build their own identity and customer base. All the experts were happy and everything met their rigorous requirements.

After five years of searching, they had finally found exactly what they were looking for. Climatically and culturally they made the giant leap from Arizona to Waiheke Island. The property deal was closed on Christmas Eve 2008 and they moved in permanently a year later. Rob undertook every job out on the vineyard and they surrounded themselves with the combined knowledge of our growing wine industry. Connie and Rob need not have worried about being accepted into the community here – they've been embraced. They've taken over management of a block nearby to produce an exquisite barrel-fermented Chardonnay. Connie's drink of choice is bubbles, and they are one of only a few sparkling wine producers on Waiheke. Their bottle-fermented vintage Blanc de Noirs is a rare blend of Cabernet Sauvignon and Merlot, made by the gifted methode traditionelle winemaker Daniel Le Brun of No.1 Family Estate in Marlborough. All other wines are made by David Evans at Passage Rock from vines that Peacock Sky manage fully. The economics of building a winery were not feasible and Connie and Rob feel privileged to work with David.

Peacock Sky is one of the highest planted vineyards on Waiheke. The steeply planted vines are well exposed to the sun and receive vitally cooling, cleansing breezes. At the top of the vineyard, panoramic views await you. Swivel your gaze in full circle, from the Coromandel Peninsula, to the waters of Onetangi Beach, around to the city and its piercing Sky Tower and on around Waiheke's southern coast. The fifty wild peacocks living on the nearby ridge, and their plumage's similarity to the palette of Waiheke skies, gave Connie and Rob their business name. The Monarch butterfly travels all the way from Montreal to Mexico and is symbolic of rebirth, new life and new beginnings. It adorns their label representing their own colourful journey and transformation.

Building projects on Waiheke have long been famous for their extended timelines and it was no different for Connie and Rob. Their anticipated opening of January 1st 2010 became April 1st 2010. From the outset, Connie wanted a point of difference for her wine tasting room. Peacock Sky is a destination requiring forward planning, a vehicle or an energetic sense of adventure. She was keenly aware that they had to provide an unforgettable experience. Her Italian heritage and passion for food inspired her decision to offer a degustation wine tasting, each wine paired with a tasty morsel of food. On opening day, they invited all the local tour operators, business owners, vineyard owners – everyone they could think of – to discover what was on offer.

To further build momentum and friendships, they started inviting people to lavishly self-catered five course dinner parties. Guests were enamoured of Connie's delicious pairings and begged for larger platters of food. Soon that wasn't enough for her appreciative audience, so Connie began serving a-la-carte dishes. Each season visitors would ask her to add a little bit more to the menu. Nowadays everything is made on site including breads, chutneys and chocolate truffles: Connie's promise to herself not to spend long hours in the kitchen has indeed been forgotten.

Requests were made for evening dining, but Connie and Rob have absolutely no plans to extend opening hours for dinner. They do host private evening functions, but only a manageable amount per month. In recent years Connie has added a cooking school dimension to the business, to give small groups the delightful experience of cooking and enjoying Peacock Sky recipes and wines in an intimate and cosy environment. It's a great way to invigorate the slower winter months that we all experience in hospitality on the island.

Connie and Rob have enhanced the wine-tasting experience on Waiheke and are keen advocates for our island of wine to offer an array of excellent and diverse experiences.

They have instigated a 'repeat offenders' club: members are people for whom one visit was never going to be enough – they return again and again to introduce friends to their Peacock Sky family, knowing they will receive a warm welcome and a completely home-made experience every single time.

ANTONIO CRISCI

ANTONIO CRISCI
PODERI CRISCI

—

WINE AND LA DOLCE VITA

Antonio Crisci was born in Naples – "the north of the south of Italy", under the gaze of Vesuvius, guarded by the islands of Capri and Ischia. Swimming, sailing and seafood in the shadow of a volcano: you can understand why Antonio feels at home here in Auckland and on Waiheke Island, his South Pacific Isola d'Ischia. At the same time, he's been infusing the culture of his first home into his new one – recreating la dolce vita down-under.

He is one of three children born into a family where fashion and food have been the trades practised for generations, and Antonio absorbed and embraced both. On his father's side is fashion and, as a boy still in short pants, Antonio went to work with his father in the industry. He started selling fabrics, then trousers and shirts, eventually specialising in the women's sector, doing everything from production of clothing to sales, exporting and importing. His family continue to have fashion business interests in Italy.

On his mother's side was food. Antonio's grandmother owned a restaurant and was a wine merchant. Antonio's family keenly observed the Italian tradition of spending hours on Sundays gathered around the table socialising and eating – and as a boy, Antonio spent Sunday mornings in the kitchen with his mother and sisters preparing the long lunch. As each of the many rich dishes to fill the lasagne were prepared, he'd jump in to make himself a sandwich bursting with every delicious item. At this young age, he wanted to taste and to understand how each part contributed to the whole dish before it was delivered to the dining table.

Following his teenage years of business training with his father in the fashion industry, Antonio travelled to France to work on a vineyard. Small but significant experiences imprinted themselves into his food memories. He couldn't believe his eyes when, at 7am, everyone pulled out their baguettes and salami, washing them down with red wine instead of coffee. By day two, he sat alongside them, tearing chunks of bread and meat, sipping wine. Eager to work with food, after years practising in the kitchens of his grandmother and his mother, he opened his first restaurant in Naples when he was just 19 years old.

Life raced by; for the next decade, he spent long hours as a chef while maintaining his interest and involvement in fashion.

Then, in his mid thirties, he and a friend left Italy to realise their long-held dream of sailing around the world. From time to time, other friends climbed aboard, but mostly it was just the two of them. They were very lucky with the weather, only once having trouble navigating their departure from Pitcairn Island, spending five days battling strong winds.

After a couple of years at sea, Antonio needed to plant his feet back on terra firma. Tonga and Fiji lay on the horizon but did not appeal for a longer stop-over. He arrived in New Zealand, in the Bay of Islands, where he resided for five months, eventually making his way down to Auckland. He got a job as a chef and the restaurant sponsored him to continue working until his work visa expired. He had to leave, but these islands at the bottom of the world had won his heart, so he applied for residency, then sailed back to Europe, to Italy, to Naples, to family. He received his life-changing letter confirming that residency had been approved and, in 1993, he arrived back in Auckland.

Toto Restaurant was opened by Antonio that year. In an ornate old building with a distinctly European sensibility on Nelson Street in the city, Antonio injected Italian food and opera into the Auckland dining scene. Toto's became an institution – but like all successful projects, it took a bit of navigation. Antonio presumed that lunch service would be the core business, catering to corporate clientele, so the restaurant offered a full lunch menu while by night, he served one four-course menu for $35. He wanted to establish regular customers, so each night was reserved for a particular type of dish – lamb shank was followed by fish, another night might feature veal. The menu was always different, always revolving.

Word spread about the delicious food. Guests would venture across the Harbour Bridge to taste the four-cheese gnocchi their colleagues had been raving about... only to find that it wasn't on the menu that night. Antonio was questioned – how dare he not have the gnocchi on the menu every night! So he changed his game plan to adapt to our slowly-growing understanding of particular European traditions.

The first chefs he employed had no idea what an eggplant was, nor what to do with it. Hearts were finally discovered amidst the overgrown artichokes delivered. Provelone was unheard of. He remembers cooking his first fresh pastas in a wok, then converting a deep-fryer into a pasta cooker, before importing one from Italy. In those early days, he had to mix different flours and semolina to create the perfect end result for his pasta.

Our coffee culture was fledgling in those days. Antonio was looking for a blend to remind him of Naples where the coffee is rich and quite different to elsewhere in Italy. It is, says Antonio, because of the water, which also influences the flavour and texture of Napolitano pastas and pizzas. He got talking to a young roaster called Michael Allpress and, together, they produced a blend especially for Toto.

In 1996, Antonio opened Non Solo Pizza in Parnell – a sister restaurant to Toto. Tired of not being able to source great Italian bread for his restaurants, around 2003 he opened Il Forno Italian bakery in Ponsonby. He's the kind of guy who, if he can't find it, then he gets on and produces it himself.

The restaurant kitchen is a very particular environment. Tensions run deep, stress levels rise as the pace gallops along during service. Music can make or break the collective pressure. What to play is vitally important, and Antonio decided that they would only listen to opera. One night a regular customer asked if his dining companion could serenade the dwindling dinner crowds and Antonio agreed. The performance was incredible, holding everyone spellbound with surprise. Antonio engaged him to sing two nights every week until he left our shores, destined for the limelight. Now world famous, Simon O'Neill is regarded as one of the finest heldentenors on the international stage.

When he arrived in New Zealand, Antonio didn't drink our local wines for the first few years. They were extremely different to what his palate was accustomed to – either too sweet or too fruity and they didn't complement his Italian dishes. He imported Italian wines for his restaurants, but slowly New Zealand producers also made their way onto his wine lists. One of the first red wines he recalls was Peninsula Estate – a Bordeaux blend produced on Waiheke Island and made by Chris Lush.

Antonio started thinking about producing his own wines in a style more accustomed to suit his food. He began looking for a vineyard and friends Basil and Julie Christensen invited him over to their vineyard in Awaawaroa Bay on Waiheke Island. They were growing Bordeaux varieties and producing wines made by Herb Friedli under the Christensen Estate label. It was a simple set up – a packing shed doubled as a place to stay overnight, and a tractor shed stored the vineyard machinery. The construction of a stone cellar was underway. They discussed planting some white wine varieties to appeal to the summer crowds and Antonio suggested developing a food concept for them.

In Italy, agriturismo – farm tourism – has paved the way for visitors to stay on family-owned farms and experience first-hand the daily and seasonal routines. Antonio had always wanted to create a similar situation for his family – a place to grow vegetables, and make wine, cheese and olive oil. To be self-sufficient and to share these home-grown products and experiences with family and visitors alike. He had narrowed his search down to two vineyards, one on Waiheke and the other south of Auckland city, in Clevedon. Out of the blue, Basil and Julie announced that they were moving to France and asked Antonio if he wanted first option to buy Christensen Estate.

In March 2008 he purchased the vineyard, renaming it Poderi Crisci. He began building his agriturismo dream – a restaurant with gardens to supply the kitchen; a home for his family; and completion of the cellar. He and his staff planted additional Italian grape varieties and created wines to match his menus, and the restaurant opened in 2010. In those early days, Antonio wondered whether visitors would venture all the way out to Awaararoa Bay, but he has grown to appreciate the distance – of being a destination, where visitors plan to spend the day, far away from the busy crowds.

It has certainly not been easy. His five-year plan became a 10-year plan. He sold Toto Restaurant to his nephew in the early 2010s to allow him the 50-60 hour weeks required in the kitchen at Poderi Crisci. His partner Vivienne Farnell has been an invaluable asset in all restaurants - maitre'd extraordinaire – while maintaining the routine of family life that restaurant hours compete with. The family live in town, and Antonio commutes between NSP and Poderi Crisci. Now with permanent Italian chefs in his Waiheke kitchen, Antonio enjoys more precious time dedicated to his family.

Much has changed since his early days in New Zealand. As I walked into Poderi Crisci in the morning, I paused to watch fresh pasta being made on the marble bench-top, and listened to the melodious Italian conversation in the open kitchen. On special occasions, live opera resonates throughout the wine cellar that serves as a private dining room. Walk through the garden with a glass of estate-grown Arneis or Montepulciano, then step into the restaurant to enjoy a long lunch. Time will slow, spiced with conversation, while you embrace fine Italian tradition long into the fading light.

RACHAEL CARTER

—

RACHAEL CARTER
SOHO

—

WINE AND CLOSURE

When Rachael Carter's male business partners on the other side of the world wouldn't listen to her suggestions to refine the screw caps for wine bottles they were manufacturing, she sent them emails from her father's email account so they'd listen to her, believing that she was her father. When they told her she wasn't capable of being the General Manager of the new screw cap manufacturing plant they established in Auckland, unbeknownst to them, she quietly stepped into the role anyway. Within three years she had gained 75% of the screw cap market and when she sold the company to start Soho Wines in 2009, she had 97% market share. Rachael was just 37 years old.

There is a competitive edge and an eye for opportunity in Rachael's family. Her father John made his own pole vault and became a New Zealand champion. Her brother Mark was an All Black and they breed blueblood racehorses. Rachael was playing tennis against her older brothers and father from the age of five. As captain of her high school hockey, squash and tennis teams, she had the focus and commitment necessary to meet a challenge head on. The family have individual and shared business interests, and share all the profits and losses. When one wins, they all win and when one fails, they all take the hit. It's a great model where they protect and encourage each other's interests.

Rachael, Mark and John Jnr grew up in Oratia, West Auckland on a seven acre farmlet. Their mother's Croatian family had owned an orchard there, and Peggy and John Snr had many friends in the local wine industry. John took an interest in the wine industry in the early 1970s: with $1000 in the bank, Brewers and Winemakers Supplies Limited began trading from the garage at home. He imported and sold corks, barrels, yeasts, filtration, enzymes and machinery to the wine industry and had the entire market to himself for a long while. When competition did finally arrive in the mid 1990s, John asked Rachael to join him to win back custom.

At the time, Rachael was only one year into a marketing degree, which covered advertising, public relations, and – particularly of interest to Rachael – how to build a brand. She understood how vital that is to succeed. So her first priority was to employ a consultant to help her rebrand her father's business. Carter and Associates was born, with Rachael working in sales, and her brother John Jnr in production. Their father was a great mentor and Rachael learnt the value of putting in long hours. Of course, working with family can be fraught, so at times they argued, and Rachael grins when remembering the many times she resigned.

When she was 23, Rachael took a temporary break from working with the family, travelling for a year with two friends. They made their way through Europe, South Africa and America. Invitations were extended from the international suppliers whom Carter and Associates had always invited into their home. The girls spent two weeks staying in an apartment in Bordeaux owned by the François Frère family who supplied them with wine barrels. There they were led through stunning Chateaux, tasting wine from barrels in magnificent halls. In Spain, they spent the weekend with the family who supplied them tin caps. Strong bonds were formed and these friendships still flourish today.

After several years, Australian brewers Foster's Group approached the Carters with an offer for the family business they couldn't refuse. With a five-year restriction of trade, the family continued working with Fosters to ensure continuity of relationships with buyers and international suppliers.

Rachael started managing John's remaining business TOPenz, which manufactured tin capsule tops for wine bottles with an Italian partner. Their old competitor introduced Stelvin, a new type of wine bottle closure, to New Zealand, and was devouring the market. But Rachael had done her research on screw caps, and could see ample opportunity. Screwcap wine bottle closure had crept into the market early in the millennium, following problems securing quality corks. The timing was perfect for Rachael to seize the market with a superior product. Meanwhile, Fosters weren't happy with the quality of caps from Italy, so Rachael took over the Global Cap Agency from Fosters, incorporating it into TOPenz. Much to John's surprise, without a single customer, she set out to prove that she could make it work.

Rachael's challenge was to develop a product that had the quality to compete against Stelvin – without the support of her business partners. The Italians didn't like dealing with Rachael – they preferred to deal with John. She was smart enough to know that she'd never get their buy-in to any of her recommendations. So she worked directly with technicians in New Zealand who implemented her design ideas and together they produced a state-of-the-art screw cap. Rachael needed the factory in Italy to fine-tune production quality to her exacting standards. She outlined the design specifications necessary and found ways of getting the Italians to listen. Her cousin Brett spent three months in Italy monitoring production. Her screwcaps triumphed, and a manufacturing plant was built in Auckland to keep up with demand. Within three years she had 75% of the market share.

Her father was immensely proud – but, separately, his health was suffering due to a few investments that caused him much anxiety. Rachael wanted to understand how she might be able to help turn these problem investments around. His major concerns were vineyards he had bought in Marlborough and on Waiheke to supply fruit on contract. The new winery that had agreed to buy the majority of his grapes could no longer commit and pulled out, leaving John with vineyards producing high quality grapes but no buyers.

Ultimately, John's problem investments were to become another family success story with Rachael at the helm. The solution to Rachael was unnervingly simple. She'd always loved wine and wine was a natural part of daily life for the family. They had been committed to the wine industry for over 40 years and had a broad spectrum of customers and friends in New Zealand and around the world. It's a supportive community, peopled with salt of the earth characters, with vision and loyalty. Rachael had worked amongst them her entire adult life, nurturing a long and quietly-held ambition to manage her own wine brand. After years of battling tirelessly in the male-dominated domain of screw caps, she had conquered the market and was absolutely ready to leave it to the Italians. She sold TOPenz to the Italians and, without so much as a backward glance, stepped straight into a new life and a new business in the wine world.

Rachael had once planned to study acting in New York, and so she named her label "Soho" after the New York quarter. It stands for everything Rachael loves in life – art, fashion, creativity, entertainment – as well as evoking the many wine bars she'd enjoyed while staying there. She built her brand to reflect the people drinking at those bars. It had to be sexy, classy, bespoke – a quality wine experience with a big splash of delight and style. She creates all the marketing and branding for Soho, combining what she loves to do with what she's really good at.

While still in the planning stages of Soho Wines in 2009, Rachael fell pregnant. She continued rearing her infant business throughout nine months in and out of hospital with complications. On what she calls the best night of her life, she gave birth to Maren, her daughter, after whom the Soho

Riesling is named. Rachael is proud of her achievements, but it's Maren who is the absolute highlight of her life. Motherhood is her priority and her working life is shaped around it.

Soho has been a challenging business but managing TOPenz – now Enoplastic NZ – taught her many valuable lessons. Above all, as boss, she will always look after her staff and treat them with respect. She credits her team for Soho's many medal successes, growing sales and 22 export markets.

From the outset, Rachael's vision was to invest in quality people, grapes and wine regions, but not to invest in capital-intensive wineries. Soho make their wines in existing wineries: a win-win for both parties. Grant Taylor from Valli produces and makes their Central Otago Pinot Noir. Dave Clouston from Two Rivers makes Soho's Marlborough wines and oversees management of Soho's Marlborough vineyards plus a group of ten small family-owned grape growers providing contract fruit. James Rowan is the wine maker for their Waiheke Carter vineyard, home to the extended Carter family when busy schedules allow.

The Soho Auckland team includes Amber Hatton, who has been with Soho from the get-go and is charged with On Premise and Export Sales. Amber travels to Soho's international markets and trade shows regularly in lieu of Rachael. She's the very essence of the Soho brand – a stylish ambassador and savvy business woman, with a great mentor in Rachael. With an innate sense of leadership and sales, combined with razor-sharp business acumen and creative branding, Rachael Carter is building another truly successful business.

What I haven't told you about Rachael is how much fun she is. She adores music and talked her way into riding alongside Mick Jagger and Keith Richards to a party after their Auckland concert in 1995. The Soho Black Collection Pinot Gris, unsurprisingly, is named Jagger. But life's quieter now that she's a Mum and at the end of the day you'll find her curled up on the couch watching movies. She loves antique cars and her ideal day would be cruising in an open top convertible 1967 230SL Mercedes with Maren, Bob her dog and the music blasting. Rachael – get behind that wheel and cruise for a while – it's totally justified.

CHRISTINA AND CRAIG THOMAS

—

CHRISTINA AND CRAIG THOMAS
BATCH WINERY

—

COOL CLIMATE WINE

Craig and Christina Thomas didn't grow their first grapes on Waiheke; they grew them on another island, in far away, cold Michigan. It's hard to ripen grapes in that part of the world. "You don't set out to make world-class wines", Craig tells me. But in planting the vineyard, valuable practical challenges replaced theoretical guidelines. It wasn't just vines they were planting – it was their future.

Craig had a keen interest in vines and winemaking from a young age; even after he had completed his studies to become an audiologist, he continued to study viticultural and winemaking textbooks from UC Davis. He and Christina planted their "practice" vineyard on his family property on Michigan's Sugar Island in 2004. On land that had once been one of five potential sites for the United Nations Offices (now based in New York), the couple planted 2500 hybrid vines in an amphitheatre. On weekends they would drive for hours and take the ferry across the border from Canada to the island in the US to work on the vineyard.

They both like to keep busy, but eventually their overloaded schedules filled with work, children and the vineyard propelled them to change their priorities. Business had been brisk: Craig opened his first audiology clinic in 1991, and growth was rapid. His family stepped in to help manage a group of clinics in Ontario – Craig, his parents and his sister Rae travelled between all the offices to provide a family presence in the company. Christina is a hearing aid specialist and they met when Craig's company acquired the clinic her mother worked in. Together they had concentrated on strengthening their growing business, putting in long hours over many years. Now they resolved to find somewhere in the world to grow grapes and make world-class wine. Somewhere with opposite seasons to Ontario in Canada so that they could avoid the depths of winter.

So they travelled. Whenever they had a holiday it was well researched with the intention of discovering the right locale. Through North America, Europe, Argentina and Chile, they tasted beautiful wines and visited stunning locations, but nothing spoke of home. There was one particular quality wine-producing country in the world they hadn't checked out – and it was one that Christina had been drawn to since high school. Finally Christina decided to visit New Zealand.

Again, they did their research and travelled to New Zealand to look at a specific vineyard for sale in the Gibbston Valley in Central Otago. Christina booked them a place to relax for a few days after the long flight, before they drove all the way down south. That place was The Boatshed on Waiheke Island. They arrived early morning in downtown Auckland in the pouring rain. Craig felt relatively refreshed and not particularly enthused about the terrible weather. He was happy to forego Waiheke and start out straight away on the great drive south. But Christina had researched Waiheke and was not going to miss the opportunity.

They boarded the ferry, arriving well before check-in time at The Boatshed. David Scott settled them on the couch with tea. Lying on the table was a real estate magazine, so they flicked through it. Two places caught their eye – one was a vineyard in receivership, the other a house. They decided to have a look at the vineyard, previously known as Ridgeview. Despite continuing rain, the site that greeted them was breathtaking. They walked through the restaurant, which opened out onto a large flat lawn. An endless vista stretched far and wide, finding its way back across the Hauraki Gulf to Auckland.

They set off to view the house, only to discover that, to their surprise, it was next door to the vineyard, on a differently named road. Returning to the vineyard later that day after the rains had gone, they sat mesmerised, watching the sun drop into the horizon, as the red glow of the sky turned to black, revealing the twinkling lights from the city. They sat entranced with the entire experience. The drive back to the Boatshed was one of quiet contemplation, perhaps even possibility, but neither of them dared to speak of it at that stage.

They drove through New Zealand, touring through wine country all the way south to Gibbston Valley. They compared everything they saw to Waiheke. Nothing did compare, so they changed their itinerary to allow for another three days on Waiheke before flying home. They nearly didn't make it back onto the island. They had no idea that the last car ferry on a Friday back then was at 6pm and that it was usually fully booked. They squeezed on board and once again they wound their way up the ridge to revisit the vineyard that had been haunting their journey.

That weekend they drove around the island – each new turn presenting another stunning view. Nothing they had seen elsewhere in the world had compelled them to act. But Waiheke had what they were looking for – a beautiful place to live with potential for the production of world-class wines. The community was colourful and diverse. They wanted their entire family to become a part of the project, so it had to be a destination both eagerly and easily reached. It was to become a second home for generations to come. "We wanted to leave a way of life and a connection to a special place on the planet", says Craig.

They were keenly aware that Waiheke was an opportunity that may never present itself again. This contemplative couple are not spontaneous people and would never make rash decisions. But the morning they were due to fly home to Canada, they were signing the papers to purchase. As we talk and watch their sons playing together outside, it still amazes them that the moment may easily have passed them by. What if they had never gotten onto that ferry in the pouring rain in 2009 and just carried on to Central Otago?

On their vineyard in Michigan they could not make great wine, but they grew to understand the process by working on the vineyard and in the winery. There were big decisions to be made on Waiheke. They gave a lot of thought to the kind of winery they wanted: one of casual sophistication, encouraging relaxed informality. Children are always welcome, so too are your dogs. The winery even offers a free pick-up and drop-off service from the ferry with their "Batch Bus", providing wonderful commentary from the long-term local driver.

As owners of a large business, they're very aware that people are the most significant factor in your success. Hiring people smarter than you are is key for Craig and Christina. Daniel Struckman is their winemaker, with previous experience in Germany, America and New Zealand. He came with the purchase and they recognised his potential immediately. As they are absentee owners for long periods, Daniel has been instrumental to their success. They keenly involve themselves in the winery, but they leave the final decisions to him and he is a superb advocate for their wines in the public domain. Ivan Aleish is their viticulturist, who they describe as a quietly focussed guy preferring to remain in the vineyard away from the limelight. But when he's prodded to speak in public, his audience leans in, captivated by his passion.

A lot of work has been done on the vineyard, pulling out inappropriate varieties and replanting. Batch Winery specialises in Chardonnay and Syrah with smaller plantings of Pinot Gris, Flora, and Riesling. Unusually for Waiheke, they produce a spicy aromatic white from Flora, a variety they inherited from the original owners. They were planning a garden at the entrance but decided that it would be a better idea to plant vines instead. They discussed what might work on the cooler site and selected Riesling, so nowadays, visitors enter the property through a welcome driveway of vines. In 2012, their beautifully sculptural winery was built, with capacity beyond their needs. The idea was to build economic success by operating on many levels. They process the fruit of other vineyards and have installed one of the few bottling lines on the island.

Once again, the family have stepped in to help. Craig's sister Rae, her husband Michael and their mother Fran initially hosted the summer crowds at Batch until the renovated restaurant was ready to open. Fran's husband Reg would happily settle in for the summer. In recent years, Christina has worked the bar, the Front of House and stepped into the kitchen to cover for a chef on holiday. Her passion is food and she would love to work in the kitchen, but it's in the bar that she can listen to conversation, interact and learn about what's working well and what needs attention. She wants to know how everything works in real time so that she can practically contribute to decision-making.

There are three collections of wines: Batch wines are a series of regional wines produced from grapes grown around New Zealand. The 'Fizz' collection is produced on-site and made from grapes grown around New Zealand, in the Charmat method – a rarity in New Zealand. 'Thomas & Sons' are their premium wines made from estate-grown grapes. 'Legacy' is a label reserved for exceptional vintages of estate wines.

And it is a legacy that Craig and Christina want their children and grandchildren to appreciate far into the future. As Craig said, "we are planting a tree that we are not going to be sitting in the shade of – it's for the next generation, if they are interested". Batch Winery is so beautiful he's certain they'll embrace their legacy. And if you've been there, you'd absolutely agree.

NEILL CULLEY

—

NEILL CULLEY
GOLDIE ESTATE

—

WINE AND INDUSTRY

Neill Culley has had a rich and varied life in our domestic wine industry, from wine making to wine marketing to wine teaching, working for Coopers Creek and Babich Wines on the mainland; and on Waiheke, for Cable Bay and then Goldie Estate, as the University of Auckland Wine Programme Director. None of this would have happened if his supervisor in postgraduate marine biology hadn't taken extended leave and told Neill he needed to take a year out to "grow up". It's not that Neill was doing anything out of the ordinary for a young student – just playing rugby and drinking beer. But that year out has now extended to three decades; marine biology's loss is wine science's gain.

Neill had actually grown up in Coromandel and Auckland, the eldest of three children. As a student, he had a part-time job at Coopers Creek vineyard where he worked with Randy Weaver and Andrew Hendry. Randy and Andrew suggested that Neill do a post-graduate diploma in wine science rather than take time out, so he spent the next two years at the prestigious Roseworthy College in Adelaide while flatting with fellow New Zealand winemaker Kim Crawford. At that time, Roseworthy had a vineyard attached to the university, so students were able to work on the vineyard and in the winery, as well as attending lectures and lab sessions. Neill loved it – decades later he now creates similar experiences for his own students.

After graduating, Neill worked for Yalumba in Angaston, located in the Barossa Valley. But Randy and Andrew got in touch with Neill to ask him to come back and set up a boutique beer brewery in Auckland. He jumped at the chance and had a great time making craft beers in the very early days of this industry. They won awards, made lots of beer, but no money. The receivers came in and the business was sold. Neill had to call the receiver every day to report how much money they'd spent. Later on he called her because he wanted to – it must be a rare man who seeks the company of his receiver. But Denise is an exceptional woman, and those phone calls and meetings bloomed into a relationship, marriage and two children.

Neill then stepped in as assistant winemaker at historic Babich Wines in Henderson – a pioneer wine company now a century old. That temporary job turned into a fascinating twelve years. Neill's experiences at the brewery had spiked his interest in understanding how to make a business work, and so when Michael Cooper left his role as Babich marketing manager to pursue his acclaimed career in wine writing, Neill stepped into the role. He combined it with study toward his MBA, as well as work in the vineyard and winery. Then, Peter Babich, who was very involved with the Wine Institute of New Zealand, suggested Neill work on the Wine Guild committee, finding international trade opportunities for New Zealand wine producers. It was a rare marketing opportunity, and Neill travelled with a New Zealand wine delegation to London where they hosted tastings at the Wine Trade Fair. "The punishment for making wine is selling it", he says – selling's the hard part.

Neill was also travelling widely through France and America to promote Babich wines. He gained an accurate understanding of how New Zealand wines compared internationally at a time when not many local winemakers had that insight. He worked a vintage in Napa, California, expanding his wine-making knowledge. Back home, Joe Babich stepped down from chief winemaking duties and Neill stepped into his shoes for the following seven years. Babich Wines had grown their vineyards and production into Gisborne, Hawkes Bay and Marlborough. Export volumes rose 100-fold from just 800 cases to approximately 80,000 during Neill's tenure.

Armed with a very wide range of experience and contacts, Neill was ready to branch out and plant vineyards of his own. A small group of enthusiastic business partners joined him to establish Cable Bay Vineyards on Waiheke Island. Planting began in 1998 – Neill was still working at Babich and he would commute to Henderson to work, then back to Waiheke to plant vines. Denise's support was invaluable. She helped with planting and joined Cable Bay to work in accounts and finances.

Neill became actively involved in the Waiheke Winegrowers Association, encouraging peer review and blind tastings to address quality issues. He'd bring Joe Babich over to review the wines, knowing that Waiheke producers had to hear how their wines rated and, in some cases, they would have to accept that they must lift their game. And they have.

In 2001, Neill left Babich to work fulltime at Cable Bay. They initiated a new model, leasing land around the island and transforming it into vineyards. Cable Bay's first vintage was in 2002 and, during the first few years, the wine was made, bottled and stored at Pleasant Valley Wines in Henderson. The first vintage on site at Cable Bay was 2008 and it was exciting to be in full production in the beautiful new winery.

Cable Bay was sold to Loukas and Caroline Petrou, and in 2014 Neill moved on. That's when Randy Weaver stepped back into his life once again, to suggest that he would be perfect to apply for the job Randy was retiring from: Directorship of the University of Auckland Wine Science programme. In 2015, Neill got the job.

He is committed to ensuring his students understand just how challenging boutique winemaking can be. Neill knows everyone in the industry here and many are happy to open their doors to his students. Herb Friedli teaches the basics of wine barrel maintenance. At Batch Winery, students experience how sparkling wine is produced.

The beauty of studying wine science with a vineyard and winery on site means Neill can extend a lecture on canopy management with a walk in the vineyard to demonstrate what they've just heard. Slide shots become three-dimensional reality as Neill introduces students to the work of Heinrich Storm (p172) and his team. Neill's there in the wee hours with the students during vintage, wise enough to let them lead the way, stepping in where necessary. They're lucky to learn from the wisdom and multi-dimensional industry experience of this insightful and remarkable man.

HEINRICH STORM

HEINRICH STORM
GOLDIE ESTATE

—

WINE AND PARTNERSHIP

Goldie Estate is a very unique operation. The property was partially gifted by the original owners, Kim and Jeanette Goldwater and family, to the University of Auckland in 2011. So it presents us with a very different story - one that began with Kim and Jeanette and is now in the hands of the University. Heinrich Storm is the winemaker for Goldie Estate and it's his job to make the wine, manage the brand, finances, budgets and forecasts for the operation.

Yet winemaking on the most historical vineyard on Waiheke Island is a long way from Heinrich's rugby-playing childhood in South Africa.

Heinrich left South Africa aged 19 with a scholarship to play rugby in Remuera, Auckland. He loved his new life and home – and even after he sustained an injury that disappointingly put an end to the rugby, he was eager to stay put. The only way to do so was to enter tertiary studies, so he selected a degree in hospitality at AUT to enable him to work his way towards permanent residency. Heinrich's parents didn't drink wine and he knew nothing about it. But when his wine studies lecturer Cameron Douglas – who carries the hard-won and internationally rare title of Master Sommelier – introduced Heinrich to the wine world, he became absolutely fascinated.

While studying he met Leslie Posie, an American student completing a Culinary Diploma. In 2004 following graduation, they moved up to Kerikeri in the far north to work for Ake Ake Vineyard and its new owners, John Clarke and Aynsley Quenault. Their job was to manage the restaurant, with Leslie in the kitchen and Heinrich, front of house. During the slower winter months, Heinrich worked in the winery and on the vineyard. It was the perfect opportunity for him to take a peek at that world and decide if it was for him. It wasn't long before he began a second degree, this time in viticulture and oenology by correspondence with the Eastern Institute of Technology in Hawkes Bay.

Many of our qualified winemakers and viticulturists have completed this degree, because it enables you to study while working on a vineyard, facilitating application and comprehension in real time. Financially, years of paying international fees did not allow Heinrich to engage in full-time study again, so it was the perfect way forward. At Ake Ake, his wine knowledge grew. John Clarke's family have been wine merchants for generations on Jersey in the Channel Islands, and Heinrich absorbed his passion and experience. In the Ake Ake Restaurant was a Fine Wine Shop, filled with rare bottles. In the rural far north, Heinrich tasted and travelled throughout the world of fine wine.

Once again, a weekend visit to Waiheke Island left a lasting impression and motivated the couple to move. In no time, Leslie had a job at a local restaurant and Heinrich began working with Herb Friedli at what was known as Christensen Estate. When the vineyard was sold to Antonio Crisci, he and Leslie moved to Goldwater Estate. Heinrich had grown his skills and had become a great asset. He could work everywhere: in the restaurant, wine-tasting at the cellar door, in the vineyard and in the winery.

The Goldwaters have a long-standing relationship with Flora Springs, one of Napa Valley's most historic wineries. They have welcomed many Goldwater staff in the past to work a vintage with them, and Heinrich was offered that chance too. The majority of Waiheke vineyards are no more than 15 acres. Currently Flora Springs owns 650 acres throughout the Napa Valley.

Heinrich worked across the many different sites which produce a breadth and depth of hand-crafted wines. The experience Heinrich gained has proven extremely beneficial and remains a highlight of his wine career to date.

Heinrich's mentors at Goldie were Elske-Saane Hunt, the winemaker, and Andy Black, the viticulturist. Andy had been there for 13 years and his assistant Nicki Wellington for around 6 years, so their wealth of knowledge was there to support Heinrich when he stepped in for Elske, who left to prepare for the birth of her child. Mike Wood, winemaker at Obsidian, was brought in as consultant, making the key decisions and working with Heinrich. Heinrich listened, learned, questioned, studied, practised and, by the time the transfer of Goldie Vineyard and Goldie Wines was complete, Heinrich was offered the job of fulltime winemaker and operations manager. Nicki Wellington is now vineyard manager and they are employees of the University of Auckland.

Goldie Estate has 15 acres of vines planted on 30 acres of land. They have Merlot, Chardonnay, Syrah, Cabernet Sauvignon, Cabernet Franc and Viognier growing. The Syrah has begun to reveal its potential and the intention is to focus on Syrah and Chardonnay, and to investigate the success of recently-planted international varieties on the island. It's been a long transition period, and the new branding was launched in June 2016. Wines produced on Waiheke come under the Goldie Estate label; these are small quantities of hand-crafted wines reflecting their unique terroir. A second label called Goldie Winemaker Series allows them the flexibility to produce wines from any other region in New Zealand and diversify their portfolio.

Goldie Estate's restaurant kitchen is managed by Leslie, and caters for à la carte dining, events and weddings. To alleviate the complications of the restaurant being closed for private functions, the Tasting Room houses a small delicatessen where guests can select picnic ingredients and wine before heading out into the beautiful grounds to enjoy the coastal views.

Heinrich feels blessed to be living here, part of a close-knit community, where someone is always ready to help you. He looks forward to his regular chats with Kim and Jeanette Goldwater when they visit and is really grateful that he was in the right place, at the right time, with the right people who encouraged both him and Leslie to achieve and believe in their abilities. Now it's his turn, giving back to this community, sharing his fascination with wine and love of hospitality.

CHRIS AND MIKE POLAND
ARAE VINEYARD

—

FROM OBSERVERS TO PARTICIPANTS

In 1986, as a recent Waiheke arrival, Mike Poland went to see his friend Kim Goldwater after work one night, seeking his advice about planting a vineyard. It was 1986 and the wine industry was busy pulling out excess vines, receiving rebates from the Government to do so. Kim's sage reply was extremely concise - he simply said "don't".

And that – for a couple of decades – was the end of the matter. Mike and his wife Chris put the idea to bed and focused on buying and enjoying wines, especially those being produced by Waiheke vineyards.

After 13 years of living in London, they had come over to the island with their young family in 1984, part of the early eighties population boom (the island's population grew from 3500 in 1978 to 4500 by 1986). Pharmacists, they were looking to buy a business, and they found one for sale on Waiheke. Whether they wanted it or not, they also got a change of pace. In 1984, there weren't any fast ferries to Waiheke – they didn't arrive until 1987. The old ferries plied the Hauraki Gulf and beyond – the Baroona and the Iris Moana amongst them. The trip could take two hours each way, sailings were few and workers commuting to Auckland were spectacularly committed. Sea Bee Air flew from Mechanics Bay, downtown Auckland into Surfdale in just 20 minutes. These amphibian aircrafts transported passengers and freight to Waiheke from 1962 until 1989.

But the change of pace must have suited them: they had a plan to either adjust to island life and give it their best shot for five years, but if it wasn't working for them, they'd move back to the city. They bought a house in Oneroa – an idyllic spot right on the beach. They operated the pharmacy and their kids, Matthew and Alice, went to the local school. Baby Emma arrived soon after. Five years later they decided to stay put.

The wine was part of the appeal. Mike was seriously partial to wine – he loved the mystique and the rich history of the elixir. He and Chris had belonged to a Wine Club in London and travelled through Europe, particularly France, discovering wine regions. Their earliest jaw-dropping wine experience was in the late 70's. Mike's brother-in-law invited them to lunch at The Waterside

Inn, on the bank of the Thames in Bray. In this Michelin-starred restaurant opened by the Roux brothers in 1972, they drank Puligny Montrachet – the legendary and seductive Chardonnay of Burgundy. To this day, Mike can taste that wine, so astonishing was his experience of it. And maybe somewhere in the very dark reaches of his mind, a vague notion formed, growing imperceptibly over the following three decades, nudging him to make a stunning Chardonnay of his own.

Once on Waiheke, Mike describes himself as "a bit of a groupie". He'd visit Stephen White, asking questions, helping with vintage. He'd met Kim and Jeanette Goldwater at a fundraiser, introducing himself as being a friend of Kim's son. It turned out that Kim's "son" was actually Kim's brother, so it was not the most auspicious of introductions! But a discussion about wine saved the day and they've been friends ever since. The first cheque the Goldwater's ever received for their wine was from Chris and Mike Poland. Jeanette was so delightedly baffled to actually receive money for their wine that she declared she might frame the cheque. The Polands eagerly helped out during many of the Goldwaters' vintages, embracing the hands-on comradery.

Chris and Mike's summer holidays involved extended family coming to stay. Nieces and nephews sprouted and grew and various branches of the family bought their own holiday homes on the island. One nephew in particular developed a fascination for wine and through an introduction from Mike, Sam worked a vintage at Stonyridge. He went on to complete wine studies at Lincoln University, trained as a winemaker with Villa Maria in New Zealand, worked in the Hunter Valley with Rosemount Estate and in California with Littorai Wines. He moved to the UK in 1997 as wine buyer for Marks and Spencer and in 2002 he became one of approximately 350 Masters of Wine living across 25 countries around the world to date: meet Sam Harrop MW.

Meanwhile, in 2011, Chris and Mike moved into a large open plan house that was once Te Ara Retreat in Cable Bay, Oneroa. The property came with a vineyard in the back yard that been planted in Chardonnay vines, leased by Cable Bay vineyard. Chris loved the property not so much for the house, but for the beautiful aspect of the vines marching down to meet the seascape. They could sit back and watch the vineyard being cared for, purchasing the wine made of the fruit at the Cellar Door nearby. It was as if they had their own vineyard, without the responsibility of making, marketing and selling the wine. They knew there was always a slight possibility that the lease may not be renewed, but that seemed as vague as the notion of making Chardonnay that Mike conceived three decades earlier when drinking Puligny Montrachet on the Thames in Berkshire.

Proudly, Mike and Chris followed the extraordinary career of their nephew Sam. After co-founding a bio-dynamic winery in Rousillon in the South of France, he set up a consultancy business. His knowledge of the business of wine, in tandem with the making of wine, culminated in demand as an advisor to the industry across Europe. He became Co-Chairman of the International Wine Challenge – the most influential wine competition in the world. And then, with his wife Karlie and their young family, they relocated from London to Waiheke like Chris and Mike before them. While their kids enjoy the great Kiwi childhood they'd experienced, Sam commutes to commitments around the wine world, with a formidable timetable of travel.

Sam's arrival on Waiheke coincided with the news that Cable Bay would not be renewing the lease on their vineyard. It was September 2012 and they were suddenly responsible for one acre of orphaned vines. The news came out of the blue and they had no idea what to do. Was it a gift, or was it a burden? They didn't really have time to think about it, but they did speak to Sam. He was certain they could make fantastic wine and was keen for the opportunity, so they went into business together. Mike and Chris would take care of the vineyard, and Sam would make and market the wine. Before they knew it, under the careful guidance of the talented viticulturist Matt Sloane, they were covering the vines with the old nets hauled out of the shed that were full of holes. Gone were the lazy summer days – they were up at dawn running through the vineyard, bashing away on tin pots and saucepans, trying desperately to scare the birds away from the fruit. No new-age drones or old-fashioned shot guns stalked their household. They tried in desperation to train their dog Georgie to chase the birds away, but she was staying put in the shade, observing the clatter. You really can't teach an old dog new tricks.

Chris and Mike look back now and laugh, because even though they'd helped out on friend's vineyards, they'd never had to surrender to the 24/7 responsibility of their own. As harvest drew close, family and friends were called upon to help out. It's a busy yet very cherished time of the year and everyone enjoys being involved. Mike and Chris were keen cellar hands, helping Sam where they could with the winemaking – watching, listening, absorbing as much as they could. Their inaugural 2013 vintage – of yes, Burgundian style Chardonnay – was released in 2015 and sold out within months.

Cedalion is the name of their single vineyard wines. In Greek mythology, Cedalion stood on the shoulders of the blinded Orion, following his directions to guide him towards Helios, the sun in the east, where he regained his sight. Issac Newton, upon his election to the Royal Society in London, declared that he stood on the shoulders of giants – referring to Galileo and a lineage of brilliant minds. Naming the wine "Cedalion" acknowledges all of those who have come before us, enabling us to achieve our dreams.

Their vineyard was renamed Arae to acknowledge a distinguished descendant of Chris. Jeremiah Horrocks was an English astronomer who first observed the transit of Venus in 1639. Chris turned to astronomy to try and connect the name of their home, 'Te Ara' to Jeremiah Horrocks' ground-breaking observation. She found her solution in the constellation called Ara, meaning 'altar' in Latin. One of the earliest known churches on Waiheke was in the bay below their property, Church Bay. The brightest star of the Ara constellation is Arae – hence Arae Vineyard, home to the Cedalion Single Vineyard Chardonnay.

Mike and Chris have had a very steep learning curve. They're 'old school' Waiheke folk, keen observers, participants and imbibers of local wines for many years. Their experiences have been a huge advantage since taking over the vineyard. It's hard work, but they are very glad to be involved. They've discovered a positive world of like-minded people who are working in the industry because they want to, not because they have to. Their old friend Kim Goldwater came to taste their Chardonnay before they had released it. It was a nerve-wracking moment with a touching ending. He loved the wine and, in a wonderful moment, he and Jeanette became the first to lay down the money for a case of Cedalion. With that gesture, a full circle was drawn around their combined lives on Waiheke, encompassing love, wine, friends and family.

WILD
ON

WILD
ON
WAIHEKE
EST. 2007
ww.wildonwaiheke.co.nz

KERYN AND SIMON MATTHEWS
WILD ESTATE VINEYARD

—

A TALE OF TWO ISLANDS

Wine has been a near-constant in Keryn Matthews' professional life since she was a teenager. Meanwhile, when she first met her husband, Simon Matthews, he had lived most of his life on small islands. Wine and islands: it seems natural that the couple would end up on Waiheke. Together, they own and operate Wild On Waiheke, a café and brewery situated on Wild Estate Vineyard, with archery and claybird shooting thrown into the mix.

I'll tell you straight off that they threw themselves into running the operation just three days after arriving back in New Zealand from Guernsey – it explains a lot about these two slightly mad, highly accomplished professionals. When I interviewed them they were three years into their wonderfully wild new life and hadn't had a single day off. But they've loved every minute of it, laughing off the long hours and fast pace.

Keryn grew up in Hawkes Bay, living on her parents' orchards, and ever since she can remember she worked weekends and holidays. As it wasn't cool to work for your parents, teenager Keryn went to work for her friend's parents at Crab Farm Winery, igniting a lifelong interest in wine. She worked in the Winery's cellar door and restaurant throughout her degree in business at the Eastern Institute of Technology in Hawkes Bay. Thanks to her industriousness, at just 21 years old, Keryn owned her first home for herself and her three-year-old daughter, Jemma. But, keen to trial life in a bigger city, she moved to Auckland in 2000, to a job with Glengarry Fine Wine Company as a corporate accounts manager. She was able to combine her business studies with her passion for wine, growing her knowledge over several years. Every job Keryn has had since has been in the wine industry, selling, marketing and advocating for the wine brands she works with. Yet she never imagined that she would be branding and selling her own wine one day.

Simon was born on the island of Hong Kong, before moving with his British parents to the island of Bermuda, and then finally Guernsey in the Channel Islands when he was three. He has two siblings and each of them was born on a different island.

Guernsey and Waiheke share a wonderful sense of community, although Guernsey's population is eight-fold our own. Both island communities refer to themselves as living 'on the rock'. And in both places, when you are 18, having spent your entire life 'on the rock', you feel the need to flee to a big city somewhere. Simon moved to London to study osteopathy. Upon graduation, he decided to use his degree to travel the world. The young osteopath arrived in Hawkes Bay in 1997 when he was 23, basing himself there for the next three years. At a Christmas party in Napier, when Keryn was home with family for the festive season, she and Simon met.

Keryn has had some fabulous jobs in the wine industry. She has worked with smaller New Zealand brands, like Odyssey and Mt Beautiful. She has worked for larger corporate players like Lion Nathan when they grew their fine wine portfolio with Distinguished Vineyards. Keryn was appointed Territory Manager for wine sales to restaurants and retailers throughout the lower North Island. She was offered a promotion in Auckland as Sales and Operations Manager of wines and spirits for Lion Nathan. Simon worked as a lecturer of osteopathy at AUT – building people management skills which would become useful in his then unimagined future. He also commuted to Hamilton for work as a locum.

They were both working hard, so whenever possible they would flee the city to relax on Waiheke Island. They adored the place, renting out homes in the summer for friends and family to stay. They visited Wild on Waiheke often, attending friends' birthday parties and dropping in for casual lunches. Never did they imagine they'd end up owning and running it themselves.

They decided to set up a permanent holiday base on Waiheke, aiming to buy the smallest, cheapest slice of paradise they could afford. On possibly the worst day of 2006 a real estate agent picked them up from Matiatia. Rain was hosing down and cloud cover made it impossible to see much at all. They visited steep, bush-clad sites within their budget, but the agent also took them to view a tiny two bedroom bach at the end of a track heading into Hekerua Bay. They fell in love with it instantly – and they did what so many have before them – they came, they saw, they bought. Simon had to hock his premium commuting car to do so, but they were now the proud owners of their first Waiheke property.

During the following six years they spent weekends and holidays at their bach. Even during winter they loved the quiet pace of life down at the bottom of the track – a great contrast to their demanding, people-infused jobs. The area was much less built up than it is now and they feel as if they had the best years there.

Keryn moved into another job as General Manager of Cellar Select Top Value Liquor Association in 2009, managing 70 liquor stores throughout New Zealand. Her role was very social, involving travelling widely for tastings and staff training. By 2012 she was keen to experience life outside of New Zealand: Jemma had left home and they had been working hard to pay their mortgages off. The time was right for Keryn to experience Simon's world, so they packed their bags and went to live on Guernsey.

They lived with Simon's Mum, whom they adore. They hadn't planned on working while they were there: apart from a bit of osteopathic work, Simon mostly rekindled old friendships, helping friends establish a range of small businesses. But out of curiosity, Keryn rang to find out more about a job advertisement and found herself becoming a wine buyer for Randalls, established in 1868, one of the island's two main breweries. She attended wine trade fairs around Europe as a wine buyer for the company's hotels, liquor stores, duty free stores and wholesaler supplying their many restaurants on the island. She collated the wine lists for all of their licensed premises and travelled to Bordeaux, London and Dusseldorf for the major wine trade fairs. She attended executive meetings in Champagne, staying at Maison Perrier Jouët in Épernay and visited regions, vineyards and homes of producers she'd only dreamt of in the past.

At the same time, socialising with family and friends is immensely important on Guernsey. Every Sunday they'd get together for a pub or restaurant lunch with a large group and spend the entire afternoon talking and laughing. On Friday nights they'd gather at the pub or at someone's home to share a takeaway meal. It's these cherished experiences they sorely miss from their lives on Guernsey.

Life was fantastic – except for the weather. As time went by, Keryn felt the pull towards home and her daughter Jemma. She and Simon began to wonder what they might do when they returned to New Zealand. Waiheke was definitely where they wanted to base themselves –small island community that it is. The solution was to buy a business.

Just before they had left New Zealand 18 months prior, they had heard that Wild On Waiheke was for sale and thought how wonderful that opportunity would be for someone. Someone else though - not them. But when scrolling through businesses for sale on the internet, they came across it: it was still for sale. The idea took hold. They had the combined skills – Simon had worked in hospitality throughout his studies and had managed cohorts of students. Keryn had the financial, marketing, sales and branding expertise necessary. So Keryn raced back to New Zealand to check it out. She sealed the purchase in February 2013 and they arrived at the end of July, just a few days before Wild On Waiheke was scheduled to reopen after their winter break.

Keryn and Simon sold everything they'd invested in to buy Wild On Waiheke - it simply had to work for them. The business is a brewery, a cafe with outdoor activities and Wild Estate Vineyard. The vineyard was already leased out to neighbouring Te Motu and that continued for the next few years while Keryn and Simon got their heads around managing a restaurant and a brewery. In summertime they employ up to 40 staff and feed 300 people for lunch. The brewery has remained in the safe hands of Master Brewer Alan Knight. He's been brewing on Waiheke since 1997, producing a range of craft beers, ginger beer and cider.

In May 2015 they took over full management of the vineyard, planting additional varieties to satisfy customer demand. Wild Estate Vineyard comprises seven acres planted with Syrah, Pinot Gris and Chardonnay. They also have Merlot, Malbec and Cabernet Franc for a Rosé and for their signature Bordeaux blend they've named The Wildling. In exceptional years Reserve labels are produced. Their philosophy is not one of world domination – Keryn and Simon just want a great life. They plan only to produce enough wine to supply their café – to drink in-house or take away.

Without funds to buy new equipment, they rustled up whatever they found on site and with help from the island's wine community, used parts to fix machinery. Keryn and Simon have come to value the community-within-a-community that is the Waiheke Winegrowers Association. Everyone is always ready to help, whether it is offering advice or equipment.

All their hard work, play and ambitions are shared with Jemma. She lives on the property and is capable of jumping in wherever and whenever needed. Keryn never imagined she'd be working with her daughter and, after years of corporate timetables, she is relishing her daughter's company and passion for the business. Keryn and Simon are here for the long haul to ensure that visitors can experience the relaxed ambience shaped by their wild island enthusiasm.

CARRIE MENDELL AND CAMPBELL AITKEN TANTALUS

—

UNLOCKING THE SECRETS OF WINE

There is no doubt that Waiheke Island has become a top tourism destination for all who appreciate great wine, delicious food and a memorable experience. Most of us who are fortunate to call this island home first came here for a day trip, or a weekend in which to eat, drink and swim, to shed the weight of the city and feel ourselves reconnect to the beauty of land and seascapes. So it was for Campbell Aitken and Carrie Mendell, who ended up relocating from Canada to Waiheke and buying a run-down vineyard restaurant in Onetangi. They named their business Tantalus and from the outset were intent on tantalizing us with a top-quality experience.

The original Tantalus is a figure in Greek Mythology, who is said to have stolen nectar from the Gods in heaven and given it to mortals. This is apt for the producers of delicious wines, but Tantalus also has personal meaning for Carrie and Campbell. It's the name of the mountain range in Canada where they met while working together – and where they fell in love. But the allure of the Tantalus held them under its spell – as you will discover.

Carrie's background is culinary. Her passion for food developed from her mother's cooking. While still at school she began working for her best friend's mum who was a chef in the ski resort of Whistler. She would travel from Vancouver each weekend to work on Saturday, which paid for her lift ticket to ski on Sunday, then complete her homework late into the night. Later, she established a catering company, but eager for more in-depth experience, she moved on to work in a restaurant with its own cellar housing private collections of wine which guests enjoyed while dining in the restaurant. British Colombia has substantial wine importation restrictions, making it difficult to access an extensive variety of international wines. Unfamiliar bottles were called for from the cellar, and she was often offered a tasting. Her interest in wine grew and, with it, her determination to source great wines and increase her knowledge. Little did she know that before long she would be part of a talented young team making their own wines on a small island off the coast of Auckland.

Campbell was born in New Zealand, growing up in Auckland, travelling extensively with his parents. They moved to Canada when Campbell was a teenager. When he met Carrie, he was managing and restructuring businesses in Adventure Recreation, helping them to become more effective and profitable. He lured Carrie to New Zealand for a holiday with the hope of seducing her with our gorgeous landscapes and inspiring people. Friends took them out sailing to Waiheke for the day. They anchored at Man O' War Bay, headed ashore in the dinghy, and relaxed with food and wines from the vineyard tasting room. Carrie and Campbell didn't go back to Auckland that day as planned – they were utterly bewitched. They extended their stay to a long weekend, exploring the island, enjoying the wineries and restaurants. Campbell had never visited Waiheke and was fascinated that this island paradise was just 35 minutes by ferry from New Zealand's largest city.

Campbell's plan was an outright success. Carrie was keen to move permanently to New Zealand and both were drawn to Waiheke, where they wanted to establish a tourism business. They developed a plan to present to investors, their family of food and wine enthusiasts, who were excited by the proposal and gave them the green light. Acutely aware of the enormity of the responsibility to deliver, Carrie and Campbell arrived in New Zealand in May 2013.

Tantalus Estate is situated on the main road to Onetangi Beach, with a bus stop right outside the door and a golf course opposite. It's surrounded by successful vineyard restaurants that deliver a variety of memorable experiences. This gave them great confidence. They wanted a business that was also a lifestyle – one that they could live alongside, experience and grow with, working hands-on with their expert team.

Their first step was to hire winemaker Alex Perez, and viticulturist Chris Ward. Alex and Chris work in tandem with each other, discussing every tiny detail with complete dedication. Alex has international vintage experience in wine and craft beer. He maintains the beehives and is studying towards the rarely acquired Master of Wine qualification. Chris had ten years of valuable experience working at Stonyridge as viticulturist, and has an extensive history in Hospitality and Wine Tourism. In 2013, both the vineyard and winery became fully accredited members of Sustainable Winegrowing New Zealand.

In the early days both the vineyard and the enormous building housing the winery and restaurant gave Carrie and Campbell riddles a-plenty. During those first few winter months they could not identify the grape varieties that had been planted in the past. Records were not accurate enough and many of the vines were diseased. But Chris was adamant that they could turn the entire vineyard around within a year, and full trust was placed in his plan. They took out thousands of vines, replanting with more suitable varieties and new clones. Five and a half hectares are planted in Cabernet Sauvignon, Merlot, Cabernet Franc, Malbec, Petit Verdot, Syrah, Grenache, Mourvèdre, Pinot Gris, Chardonnay and Viognier.

When I first interviewed Carrie and Campbell, the old bones of the restaurant building were undergoing major reconstruction. I walked with them through the unroofed wine-tasting, kitchen and dining rooms opening out onto the vineyard, imagining what those bare spaces would become. We edged our way downstairs to the darkened interior housing the private dining room, the brewery, the speak-easy brewers lounge and the wine cellar. Only the bell tower remained as it was, albeit furnished with a grand new bell.

I was struck by the magnitude of what they were building, their attention to detail. I felt privy to their hopes and dreams, yet I was also wary of the weight of responsibility they carried, albeit with focus, intent and bright intelligence.

When they began to describe the process of naming their business and their wines, it became very clear to me that the combined talent of this young couple is a winning formula. Their ideas and words weave together seamlessly.

Tantalus. You don't hear this word spoken aloud and you may not know it at all. It's a thing of the past. In olden times nobility would lock their spirits in the Tantalus to stop the wayward or thirsty poaching their drink. One could admire, but not obtain the contents within – very tantalizing. Carrie and Campbell were drawn to the lure, the secrecy and the desire evoked by the Tantalus. Making great wines requires patience, skill and determination in order to unlock and reveal an understanding of the nature of your land – your terroir.

When designing their wine labels, they focused on the lock-and-key mechanism of a Tantalus as a powerful symbol. The colours for the labels and the font are dark and moody and you have to move in to look closely at what is hidden in plain sight. They found a large, old lock and key originating from England while browsing in Oneroa antique store, Victorian Guilt. It was prized by the shop owner and had been fitted into her own front door for years. It lay hidden under a pile of stuff because she was reluctant to part with it, telling them it wasn't for sale. Campbell and Carrie persisted and promised to provide it with a worthy new home, attaching it onto the heavy door in the cellar guarding the wine collection.

Écluse. Évoque. Voilé. These are the names of their Reserve single-estate red wines. Their French names honour the great wine appellations in Bordeaux and the Northern Rhône whose styles and quality they aim to emulate. Écluse is a canal lock – an enclosed body of liquid, like the contents of the Tantalus. This wine is Left Bank Bordeaux in style and is Cabernet Sauvignon dominant. Évoque "calls to mind" centuries of traditional viticulture and winemaking. This wine is Right Bank Bordeaux in style and is Merlot dominant. Voilé means veiled or shrouded, tying into the concept of secrecy and desire. This Syrah pays homage to its spiritual home in the Northern Rhône. Cachette is the Reserve Chardonnay. It is produced in tiny amounts and, as its label describes, it may well hide away from the crowd, so small is the chance to acquire this finely-tuned white. Alongside the Reserve wines they have produced a second Bordeaux blend of Merlot and Cabernet Franc, and a Pinot Gris.

Campbell developed a keen appreciation of craft beer in the ten years he spent in Canada. Before he left New Zealand, craft beer was in its infancy, but by the time he returned, it was flourishing. To enhance the visitor experience they've built a state-of-the-art brewery named Alibi, providing thirst-quenching home-crafted beers and ciders.

At the time of writing, Tantalus was only something I could 'évoque'. I had tasted the beautiful wines, but the building was very much a work in progress. It has been years in the making and every tiny detail has been carefully considered. The lights are a vinous triumph. The wines have been released for appraisal, rating highly. The beers, crafted by Bernard Neate, are flowing from the taps. Thousands of native trees have been planted in the wetland below the restaurant and the native eels are returning. The beehives are churning out honey; the olives and macadamias have been processed for the kitchen.

As I complete the final edit, the restaurant and cellar door have been open for a few busy months. Many have entered into the stylish, relaxed interior, enchanted by the total experience. Carrie and Campbell are hands-on in the restaurant, the tasting room and the brewery. Years of planning have come to fruition. Any spare moments are spent walking in the vineyard with their dog Brix and their adorable cat Mowgli galloping alongside. Carrie and Campbell are deeply committed to enhancing and unlocking the many possibilities that lie ahead. Their present is undeniably fulfilling. Their future is absolutely tantalizing!

CHRONOLOGY OF WAIHEKE VINEYARDS

1978	GOLDWATER
1982	STONYRIDGE
1988	TE MOTU
1989	FENTON
1992	MUDBRICK
1993	MAN O'WAR
	MIRO
	OBSIDIAN
1994	PASSAGE ROCK
1996	TE WHAU
	KENNEDY POINT
	AWAROA
1998	CABLE BAY
2000	DESTINY BAY
2001	MICHAEL'S
2002	JURASSIC RIDGE
2003	HAY PADDOCK
2006	WOODSIDE HILL
	FRENCHMAN'S HILL
2008	PEACOCK SKY
	PODERI CRISCI
2009	SOHO
	BATCH
2011	ARAE
2013	WILD ESTATE
	TANTALUS

Mother Nature J Aloch 63.

19°25'E 2012 (1'E)
Harbour and Pilotage Limits
Thompsons Point
Hakaimango Point
(14) Nani Island
Opopoto Bay
Needle Rocks (5)
Onetangi Bay
Onetangi
Oneroa Bay
Oneroa
Palm Beach
Matiatia Bay
Te Whetumatarau Point
Blackpool
Te Miro Bay
Church Bay
Motukaha Island (16)
Surfdale
Huruhi Bay
Ostend
Airfield
WAIHEK
Cable Bay
Park Point
Te Wharau Bay
Maunganui Point
Putiki Bay
Oakura Bay
Takirau Bay
Kennedy Point
Te Whau Bay
Te Whau Point
Omiha
Rocky Bay
(18) Koi Island
Papakohatu Island (10)
Sergeant Channel
T A M A K I
S T R A
Omana Beach